McGraw-Hill
My Math

Welcome to My Math — your very own math book! You can write in it — in fact, you are encouraged to write, draw, circle, explain, and color as you explore the exciting world of mathematics. Let's get started. Grab a pencil and finish each sentence.

My name is _____ .

My favorite color is _____ .

My favorite hobby or sport is _____ .

My favorite TV program or video game is

_____ .

My favorite class is _____ .

Math, of course!

McGraw Hill Education

MHEonline.com

STEM McGraw-Hill is committed to providing instructional materials in Science, Technology, Engineering, and Mathematics (STEM) that give all students a solid foundation, one that prepares them for college and careers in the 21st century.

Send all inquiries to:
McGraw-Hill Education
8787 Orion Place
Columbus, OH 43240

ISBN: 978-0-07-668888-3 (*Volume 2*)
MHID: 0-07-668888-7

Printed in the United States of America.

4 5 6 7 8 9 LMN 20 19 18 17

Meet The Artists!

Caelyn Cochran

The Angles of Ballet I was inspired by the gesture mannequin in art class. I thought of how the body moves in different ways and how ballet dancers create different poses. *Volume 1*

Syed Amaan Rahman

Step by Step - Stairway to Learning To me, math means learning in a step-by-step process. I drew consecutive numbers and math signs marching up a staircase and into light to symbolize the process of learning math and beyond with endless possibilities. *Volume 2*

Other Finalists

Jeannine Demarzo
White Cake Recipe

Nate Mitchell
Pixel Man

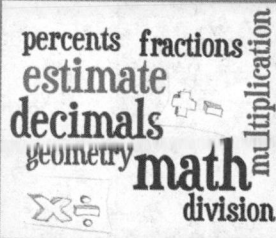

Rachel Harvey
3-D Word Collage

Dante Washington
Insect Mandala 6

Kathleen Saier
Mathematical Language of Computers

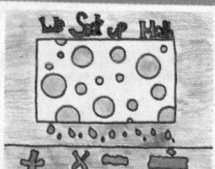

Phobe Hanscom, Hope Sternenberg, Haylee Harvey
Sponge Math

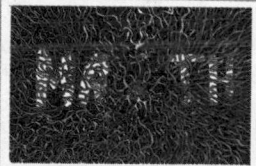

Mazi Blankenship, Abby Gunther, Zach Ginn
Mathematical Quagmire

Abigail Africa
Math is a Fact of Life

Madeline Murphy
Math Collage

Jessica Li
All the Missing Pieces

Find out more about the winners and other finalists at www.MHEonline.com.

We wish to congratulate all of the entries in the 2011 *McGraw-Hill My Math* "What Math Means To Me" cover art contest. With over 2,400 entries and more than 20,000 community votes cast, the names mentioned above represent the two winners and ten finalists for this grade.

* Please visit mhmymath.com for a complete list of students who contributed to this artwork.

iii

GO digital

it's all at
connectED.mcgraw-hill.com

Go to the Student Center for your eBook, Resources, Homework, and Messages.

Write your Username [   ] ✎   Password [   ] ✎

Get your resources online to help you in class and at home.

Vocab

Find activities for building vocabulary.

Watch

Watch animations of key concepts.

Tools

Explore concepts with virtual manipulatives.

Check

Self-assess your progress.

eHelp

Get targeted homework help.

Games

Reinforce with games and apps.

Tutor

See a teacher illustrate examples and problems.

GO mobile →

Scan this QR code with your smart phone* or visit mheonline.com/stem_apps.

*May require quick response code reader app.

Available on the App Store

Contents in Brief

Organized by Domain

Standards for Mathematical PRACTICE → *Woven Throughout*

Chapter 1

Place Value

ESSENTIAL QUESTION
How does the position of a digit in a number relate to its value?

Getting Started

Lessons and Homework

Wrap Up

Are you ready for the great outdoors?

Look for this!
Click online and you can watch videos that will help you learn the lessons.

connectED.mcgraw-hill.com

Chapter 2 Multiply Whole Numbers

ESSENTIAL QUESTION
What strategies can be used to multiply whole numbers?

Taking care of my pets!

Getting Started

Lessons and Homework

Wrap Up

connectED.mcgraw-hill.com

Chapter 3 Divide by a One-Digit Divisor

ESSENTIAL QUESTION
What strategies can be used to divide whole numbers?

Let's divide and conquer these chores!

Look for this! eHelp
Click online and you can get more help while doing your homework.

connectED.mcgraw-hill.com

ix

Chapter 4
Divide by a Two-Digit Divisor

ESSENTIAL QUESTION
What strategies can I use to divide by a two-digit divisor?

How can I divide large numbers?

Getting Started

Lessons and Homework

Wrap Up

Look for this! **Tools**
Click online and you can find tools that will help you explore concepts.

connectED.mcgraw-hill.com

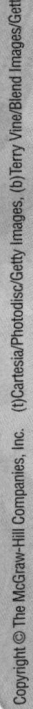

Chapter 5 Add and Subtract Decimals

ESSENTIAL QUESTION
How can I use place value and properties to add and subtract decimals?

Getting Started

Lessons and Homework

Wrap Up

Technology on the go!

connectED.mcgraw-hill.com

Chapter 6
Multiply and Divide Decimals

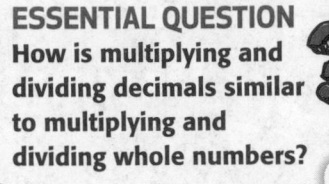

ESSENTIAL QUESTION
How is multiplying and dividing decimals similar to multiplying and dividing whole numbers?

Getting Started

Lessons and Homework

Wrap Up

Cool treats in the summer heat!

connectED.mcgraw-hill.com

Chapter 7
Expressions and Patterns

Getting Started

Lessons and Homework

Wrap Up

Look for this!
Click online and you can watch a teacher solving problems.

connectED.mcgraw-hill.com

Chapter 8

Fractions and Decimals

ESSENTIAL QUESTION
How are factors and multiples helpful in solving problems?

Getting Started

Lessons and Homework

Look for this!

Click online and you can find activities to help build your vocabulary.

Wrap Up

connectED.mcgraw-hill.com

Chapter 9 Add and Subtract Fractions

ESSENTIAL QUESTION
How can equivalent fractions help me add and subtract fractions?

Time to harness the fractions!

Getting Started

Lessons and Homework

Wrap Up

connectED.mcgraw-hill.com

Chapter 10

Multiply and Divide Fractions

What's cooking in the kitchen?

connectED.mcgraw-hill.com

Chapter 11 Measurement

ESSENTIAL QUESTION
How can I use measurement conversions to solve real-world problems?

6 gallons

Getting Started

Lessons and Homework

Look for this! **Check** ✓
Click online and you can check your progress.

Wrap Up

connectED.mcgraw-hill.com

Chapter

12 Geometry

FROM: HERE
TO: THERE

Math is an adventure!

connectED.mcgraw-hill.com

8 Fractions and Decimals

ESSENTIAL QUESTION

How are factors and multiples helpful in solving problems?

Let's Play Games and Sports!

Watch

Watch a video!

MY Standards

Number and Operations – Fractions

5.NF.3 Interpret a fraction as division of the numerator by the denominator $\left(\frac{a}{b} = a \div b\right)$. Solve word problems involving division of whole numbers leading to answers in the form of fractions or mixed numbers, e.g., by using visual fraction models or equations to represent the problem.

5.NF.5 Interpret multiplication as scaling (resizing), by:

5.NF.5b Explaining why multiplying a given number by a fraction greater than 1 results in a product greater than the given number (recognizing multiplication by whole numbers greater than 1 as a familiar case); explaining why multiplying a given number by a fraction less than 1 results in a product smaller than the given number; and relating the principle of fraction equivalence $\frac{a}{b} = \frac{n \times a}{n \times b}$ to the effect of multiplying $\frac{a}{b}$ by 1.

Number and Operations in Base Ten *This chapter also addresses these standards:*

5.NBT.5 Fluently multiply multi-digit whole numbers using the standard algorithm.

Cool! This is what I'm going to be doing!

Standards for Mathematical PRACTICE

1. Make sense of problems and persevere in solving them.
2. Reason abstractly and quantitatively.
3. Construct viable arguments and critique the reasoning of others.
4. Model with mathematics.
5. Use appropriate tools strategically.
6. Attend to precision.
7. Look for and make use of structure.
8. Look for and express regularity in repeated reasoning.

● = focused on in this chapter

Name _____

Am I Ready?

Check ← Go online to take the Readiness Quiz

Find the factors of each number.

1. 8 _____

2. 12 _____

3. 6 _____

4. 21 _____

5. 32 _____

6. 45 _____

Multiply or divide.

7. $5 \times 12 =$ _____

8. $64 \div 4 =$ _____

9. $56 \div 7 =$ _____

10. $9 \times 13 =$ _____

11. Claire brought three packages of sports drinks to the volleyball game. If each package contains 8 drinks, how many total drinks did Claire bring?

_____

Graph each number on the number line provided.

12. 6.3

13. 1.8

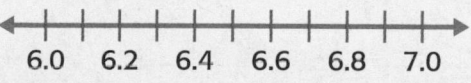

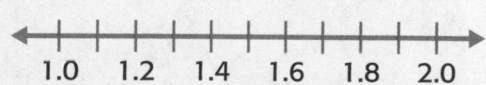

Shade the boxes to show the problems you answered correctly.

How Did I Do? ➤ | 1 | 2 | 3 | 4 | 5 | 6 | 7 | 8 | 9 | 10 | 11 | 12 | 13 |

Online Content at ⌕ connectED.mcgraw-hill.com

MY Math Words

Vocab abc

Review Vocabulary

> decimal   equivalent decimals   multiples   prime factorization

Making Connections

Use the chart below to identify the review vocabulary based on the examples provided. Then provide your own non-example. The first one is done for you.

Word	Example	Non-Example
equivalent decimals	$5.25 = 5.250$	$3.05 > 3.005$
	$2 \times 2 \times 3 \times 3 = 36$	
	twenty-one hundredths	
	3, 6, 9, 12, 15, 18, and 21 are multiples of 3.	

Explain why 0.5 and 0.50 express the same amount.

_____

_____

_____

Mathematical
PRACTICE ➡

Lesson 8–2

common factor

12: 1, 2, 3, 4, 6, 12
30: 1, 2, 3, 5, 6, 10, 15, 30

common factors of 12 and 30: 1, 2, 3, 6

Lesson 8 5

common multiple

4: 4, 8, 12, 16, 20, 24, 28, 32, 36 . . .
6: 6, 12, 18, 24, 30, 36, 42 . . .

common multiples of 4 and 6: 12, 24, 36

Lesson 8–1

denominator

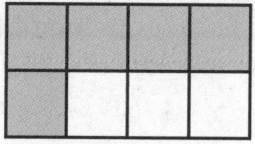

$$\frac{5}{8} = 5 \div 8$$

Lesson 8–3

equivalent fractions

$$\frac{8}{14} = \frac{4}{7}$$

Lesson 8–1

fraction

$$\frac{2}{3}$$

Lesson 8–2

greatest common factor (GCF)

12: 1, 2, 3, 4, 6, 12

30: 1, 2, 3, 5, 6, 10, 15, 30

greatest common factor: 6

Lesson 8–6

least common denominator (LCD)

$$\frac{2}{3} = \frac{2 \times 2}{3 \times 2} = \frac{4}{6}$$
$$\frac{1}{6} = \frac{1 \times 1}{6 \times 1} = \frac{1}{6}$$

} **6 is the LCD**

Lesson 8–5

least common multiple (LCM)

4: 4, 8, **12**, 16, 20, **24**, 28, 32, **36** . . .
6: 6, **12**, 18, **24**, 30, **36**, 42 . . .

least common multiple of 4 and 6: 12

Ideas for Use

- Develop categories for the words. Sort them by category. Ask another student to guess each category.

- Design a crossword puzzle. Use the definition for each word as the clues.

A whole number that is a multiple of two or more numbers.

Use a thesaurus to write a synonym and an antonym for *common*.

_____

A number that is a factor of two or more numbers.

When are factors useful?

_____

_____

Fractions that have the same value.

Give an example of two fractions that are equivalent.

_____

_____

The bottom number in a fraction. It represents the total number of equal parts.

Write a tip to help you remember which number is the denominator in a fraction.

_____

_____

The greatest of the common factors of two or more numbers.

Find the greatest common factor of 24 and 36. Show your work.

_____

A number that represents equal parts of a whole or parts of a set.

Explain how fractions represent division.

_____

_____

The least multiple, other than 0, common to sets of multiples.

How can finding the LCM of two numbers be helpful in a real-world situation?

_____

_____

The least common multiple of the denominators of two fractions.

Explain why finding the LCD is important in working with fractions.

_____

_____

MY Vocabulary Cards

 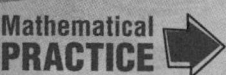

multiple

7,	14,	21,
1 × 7	2 × 7	3 × 7
28,	35,	42,...
4 × 7	5 × 7	6 × 7

numerator

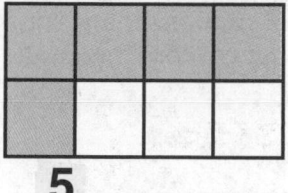

$$\frac{5}{8} = 5 \div 8$$

simplest form

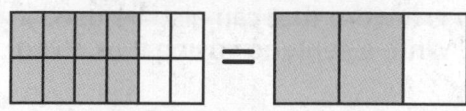

$$\frac{8 \div 4}{12 \div 4} = \frac{2}{3}$$

Ideas for Use

- Write a tally mark on each card every time you read the word in this chapter or use it in your writing. Challenge yourself to use at least 3 or 4 tally marks for each card.

- Write the names of lessons you would like to review on the front of each blank card. Write a few study tips on the back of each card.

The top number in a fraction. It tells how many of the equal parts are being used.

The Latin root *numer* means "number." Write two words that contain this root.

_____

_____

A multiple of a number is the product of that number and any whole number.

Explain why 32 is a multiple of 8. What is another multiple of 8?

_____

_____

A fraction in which the GCF of the numerator and the denominator is 1.

Form is a word that can also be used as a verb. Write a sentence using it as a verb.

_____

_____

FOLDABLES Follow the steps on the back to make your Foldable.

Fractions	Decimals	Models
$\frac{1}{2}$	0.5	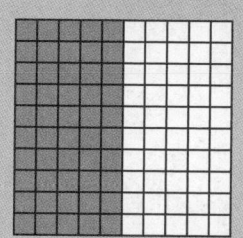

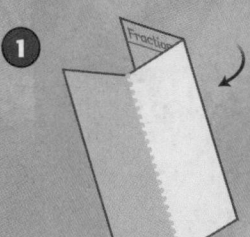

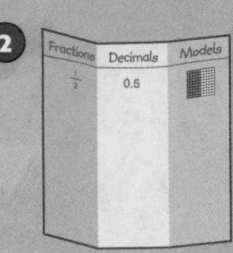

Name _____

Fractions and Division

Copyright © The McGraw-Hill Companies, Inc.   Comstock Images/Alamy

Lesson 1

ESSENTIAL QUESTION
How are factors and multiples helpful in solving problems?

A **fraction** is a number that names equal parts of a whole or parts of a set. A fraction represents division of the numerator by the denominator.

$$\text{numerator} \longrightarrow \frac{1}{3} = 1 \div 3 \longleftarrow \text{denominator}$$

The **numerator** is the number of parts represented. The **denominator** represents the number of parts in the whole.

Fast break for food!

Math in My World

Example 1

Dylan, Drake, and Jade are sharing 2 small pizzas equally after their lacrosse game. How much does each person get?

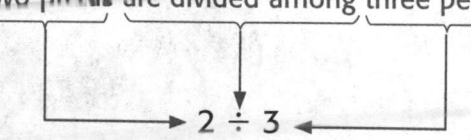

Two pizzas are divided among three people.

$$2 \div 3$$

Each person gets $\frac{2}{3}$ of a pizza.

So, $2 \div 3 = \dfrac{\square}{\square}$.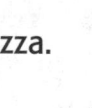

The amount of pizza each person gets is between what two whole numbers?

_____

Example 2

Ray and Bailey are sharing 3 brownies equally. How much does each person get?

Three brownies are divided between two people.

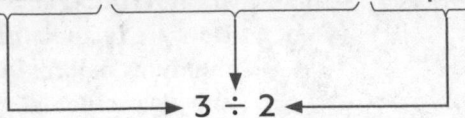

$$3 \div 2$$

Each person gets $\frac{3}{2}$ of a brownie.

It's a sweet life!

The models at the right show that each person gets 1 whole brownie. There is one brownie remaining, which can be divided equally between the two people.

Ray	Bailey	Bailey
	Ray	

Each person gets $1\frac{1}{2}$ brownies.

So, $3 \div 2 = \boxed{}\frac{\boxed{}}{\boxed{}}$.

The amount of brownies each person gets is between what two whole numbers?

_____

Talk MATH

Give an example of how a fraction represents a division situation in real life.

Guided Practice

1. Two bags of birdseed are used to fill three bird feeders. How much birdseed does each feeder use? Represent the situation using the model. Then solve.

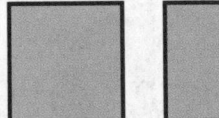

Each feeder uses $\frac{\boxed{}}{\boxed{}}$ of a bag of birdseed.

So, $2 \div 3 = \frac{\boxed{}}{\boxed{}}$.

Chapter 8 Fractions and Decimals

Independent Practice

Represent each situation using a model. Then solve.

2. Four families equally share 5 pies. How much pie will each family receive?

Each family receives $\dfrac{}{}$, or $\dfrac{}{}$, of a pie.

So, $5 \div 4 = \dfrac{}{}$.

The answer is between the whole numbers _____ and _____.

3. Six bags of soil are used to fill 5 flower pots. How much soil does each flower pot use? Between what two whole numbers does the answer lie?

Each flower pot uses $\dfrac{}{}$, or $\dfrac{}{}$, of a bag of soil.

So, $6 \div 5 = \dfrac{}{}$.

The answer is between the whole numbers _____ and _____.

4. Forty yards of fabric are used to make 9 school banners. How many yards of fabric does each banner use? Between what two whole numbers does the answer lie?

Each banner uses $\dfrac{}{}$, or $\dfrac{}{}$, yards of fabric.

So, $40 \div 9 = \dfrac{}{}$.

The answer is between the whole numbers _____ and _____.

Problem Solving

5. **Mathematical PRACTICE 5** **Use Math Tools** Demont used 4 gallons of gasoline in three days driving to work. Each day he used the same amount of gasoline. How many gallons of gasoline did he use each day?

_____

6. Suzanne made 2 gallons of punch to be divided equally among 10 people. How much of the punch did each person receive?

_____

7. The baseball team is selling 30 loaves of banana bread. Each loaf is sliced and equally divided in 12 large storage containers. If each slice is the same size, how many loaves of banana bread are in each container? Between what two whole numbers does the answer lie?

_____

_____

My Work!

HOT Problems

8. **Mathematical PRACTICE 2** **Reason** You know that if $15 \div 3 = 5$, then $5 \times 3 = 15$. If you know that $7 \div 8 = \frac{7}{8}$, what can you conclude about the product of $\frac{7}{8}$ and 8?

9. **Building on the Essential Question** How can division be represented by using a fraction?

_____

_____

_____

MY Homework

Homework Helper

Need help? connectED.mcgraw-hill.com

Two truckloads of mulch are used to cover seven playground areas. Each playground receives the same amount of mulch. How much mulch does each playground receive?

Two truckloads are divided to cover seven playgrounds.

$2 \div 7$

1	2	3	4	5	6	7

1	2	3	4	5	6	7

Each playground receives $\frac{2}{7}$ of a truckload.

So, $2 \div 7 = \frac{2}{7}$.

This number is between the whole numbers 0 and 1.

Practice

1. Three pounds of potatoes make eight equal-size servings of mashed potatoes. How many pounds of potatoes are in each serving? Represent the situation with a model. Then solve.

Each serving uses 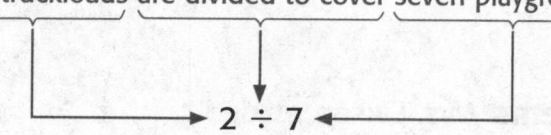 pound of potatoes.

So, $3 \div 8 = \boxed{}$.

The answer is between the whole numbers _____ and _____.

Problem Solving

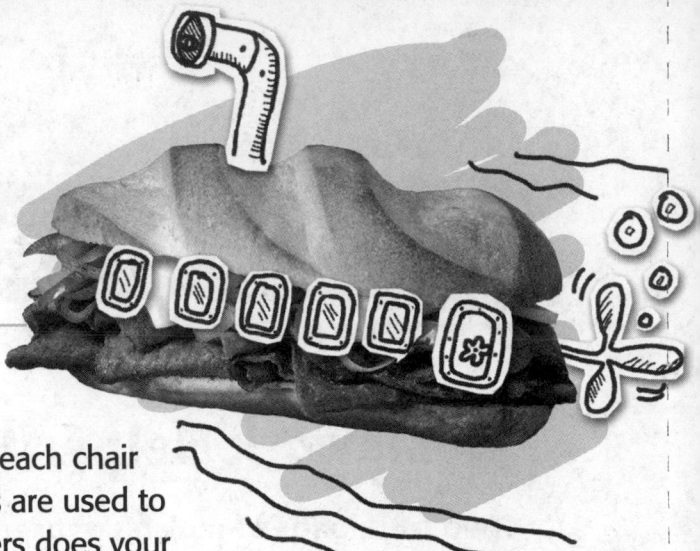

2. One large submarine sandwich is divided equally among four people. How much of the sandwich did each person receive?

_____

3. Four gallons of paint are used to paint 25 chairs. If each chair used the same amount of paint, how many gallons are used to paint each chair? Between what two whole numbers does your answer lie?

_____

4. **Mathematical**  **PRACTICE** **Make Sense of Problems** Mrs. Larsen made 12 pillows from 16 yards of the same fabric. How much fabric was used to make each pillow? Between what two whole numbers does your answer lie?

_____

Vocabulary Check

5. Fill in each blank with the correct word to complete the sentence.

The numerator is the _____ number in a fraction, while the

denominator is the _____ number in a fraction.

Test Practice

6. Elena drank 5 bottles of water over 7 volleyball practices. How much water did Elena drink each practice if she drank the same amount each time?

Ⓐ $\frac{2}{7}$ bottle

Ⓒ $\frac{5}{7}$ bottle

Ⓑ $\frac{2}{5}$ bottle

Ⓓ $\frac{7}{5}$ or $1\frac{2}{5}$ bottles

Greatest Common Factor

Lesson 2

ESSENTIAL QUESTION
How are factors and multiples helpful in solving problems?

Factors shared by two or more numbers are called **common factors**. The greatest of the common factors of two or more numbers is the **greatest common factor (GCF)** of the numbers.

 ## Math in My World  Watch  Tutor

Example 1

Sevierville Middle School arranges their sports trophies in rows in a display case. There is an equal number of trophies in each row. Each row has only one kind of trophy. What is the greatest possible number of trophies in each row?

Trophies	
Type	**Number**
Volleyball	40
Football	24
Baseball	32

Write the prime factorization to find common factors.

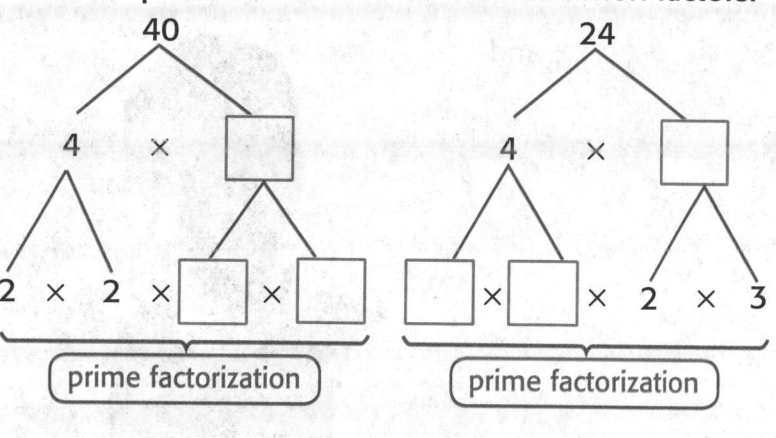

The common prime factors are 2, 2, and 2.

Multiply to find the GCF.

_____ × _____ × _____ or _____

So, the greatest number of trophies that could be placed in each row is _____.

WE'RE PROUD OF YOU!

Example 2

Find the GCF of 60 and 54.

Make an organized list of the factors for each number. Then circle the common factors.

60: 1, 2, 3, 4, 5, 6, 10, 12, 15, 20, 30, 60

54: 1, 2, 3, 6, 9, 18, 27, 54

The common factors are _____, _____, _____, and _____.

So, the greatest common factor, or GCF, of 60 and 54 is _____.

Guided Practice

Find the GCF of each set of numbers.

1. 8, 32

   8: _____

   32: _____

   The common factors are _____, _____, _____, and _____.

   So, the GCF of 8 and 32 is _____.

2. 3, 12, 18

   3: _____

   12: _____

   18: _____

   The common factors are _____ and _____.

   So, the GCF of 3, 12, and 18 is _____.

Talk MATH

Explain which method you prefer to find the GCF of two numbers.

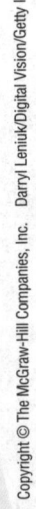

Independent Practice

Find the GCF of each set of numbers.

3. 24, 60 _____

4. 12, 18 _____

5. 18, 42 _____

6. 30, 72 _____

7. 1, 10, 14 _____

8. 14, 35, 84 _____

9. 9, 18, 42 _____

10. 16, 52, 76 _____

Problem Solving

11. Mathematical **PRACTICE** **3** **Justify Conclusions** Annika is placing photos in a scrapbook. Each page will have only one size of photo. She also wants to place the same amount of photos on each page. What is the greatest number of photos that could be on each page? Justify your response.

Scrapbooking	
photo size	photos
Large	8
Medium	12
Small	16

My Work!

_____

12. Twelve pens and 16 pencils will be placed in bags with an equal number of each item. What is the most number of bags that can be made?

_____

13. Oliver has 14 chocolate chip cookies and 21 iced cookies. Oliver gives each of his friends an equal number of each type of cookie. What is the greatest number of friends with whom he can share the cookies?

_____

HOT Problems

14. Mathematical **PRACTICE** **3** **Which One Doesn't Belong?** Circle the number that you would take away so that 8 will be the GCF of the remaining three numbers.

| 16 | 8 | 24 | 20 |

15. **Building on the Essential Question** How can you find the greatest common factor of two numbers?

_____

MY Homework

Homework Helper

Need help? connectED.mcgraw-hill.com

The table shows the amount of money Ms. Ayala made over three days selling 4-inch × 6-inch prints at an arts festival. Each print costs the same amount. What is the most each print could have cost?

Ms. Ayala's Artwork	
Day	Cost ($)
Friday	60
Saturday	144
Sunday	96

Write the prime factorization to find common factors.

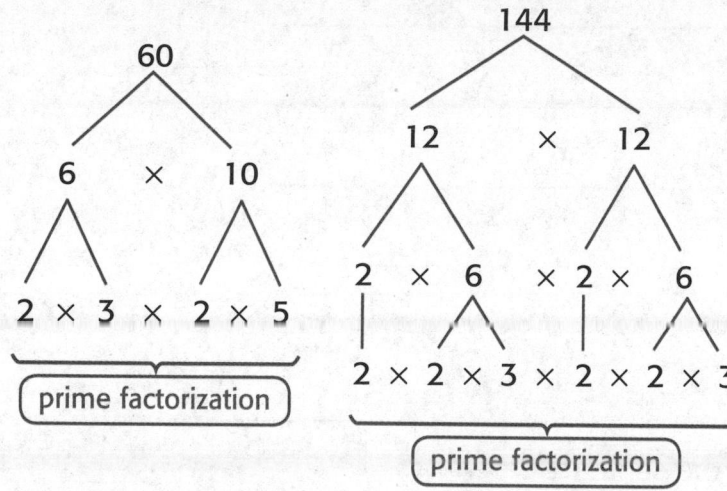

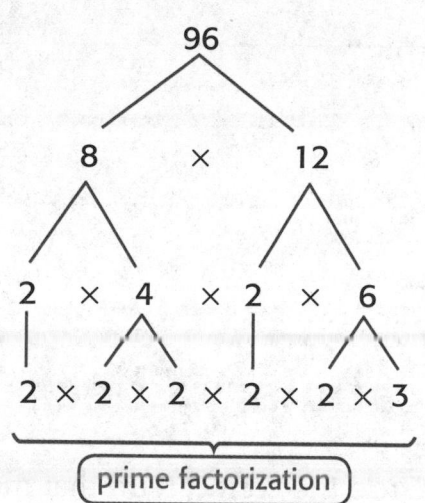

The common prime factors are 2, 2, and 3.

Multiply to find the GCF. $2 \times 2 \times 3 = 12$

So, the greatest cost of each print would be $12.

Practice

Find the GCF of each set of numbers.

1. 21, 30 _____

2. 12, 30, 72 _____

Problem Solving

I'm the leader of the pack!

3. A store sells boxes of juice in equal-size packs. Garth bought 18 boxes, Rico bought 36 boxes, and Mai bought 45 boxes. What is the greatest number of boxes in each pack? How many packs did each person buy if each box contained the greatest number of boxes possible?

_____

_____

4. Mathematical PRACTICE 3 Justify Conclusions The GCF of any two even numbers is always even. Determine whether the statement is *true* or *false*. If true, explain why. If false, give a reason.

_____

_____

_____

_____

Vocabulary Check

5. Circle the correct term that makes the sentence true.

The (greatest, least) of the common factors of two or more numbers is the (greatest, least) common factor of the numbers.

Test Practice

6. Jeremiah will share his collection with his friends so that they each receive the same number of cards. What is the greatest number of cards they will each receive?

Sports Cards	
Type	**Number**
Baseball	32
Football	24

Ⓐ 4 cards   Ⓒ 12 cards

Ⓑ 8 cards   Ⓓ 16 cards

Simplest Form

A fraction is written in **simplest form** when the GCF of the numerator and the denominator is 1. The simplest form of a fraction is one of its many equivalent fractions. **Equivalent fractions** are fractions that name the same number.

I can touch the sky!

Math in My World

Example 1

Angie has a vertical jump height of 12 inches, and Holly has a vertical jump height of 22 inches. So, Angie's vertical jump is $\frac{12}{22}$ of the vertical jump of Holly. Write the fraction in simplest form.

1 List the factors. Circle the GCF.

12: 1, 2, 3, 4, 6, 12
22: 1, 2, 11, 22

The GCF of 12 and 22 is _____.

2 Divide both the numerator and the denominator by the GCF. This results in an equivalent fraction.

$$\frac{12}{22} = \frac{12 \div 2}{22 \div 2} = \frac{\boxed{}}{\boxed{}}$$

The GCF of 6 and 11 is _____.

So, Angie's vertical jump is $\frac{\boxed{}}{\boxed{}}$ of Holly's vertical jump.

Check Use models.

Shade 12 out of 22.

Shade 6 out of 11.

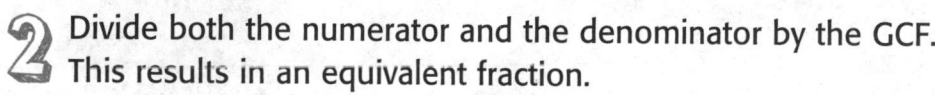

Example 2

Write $\frac{18}{30}$ in simplest form.

Divide the numerator and denominator by the same common factor. Continue dividing until the fraction is in simplest form.

$$\frac{18}{30} = \frac{18 \div \boxed{2}}{30 \div \boxed{2}} = \frac{\boxed{}}{\boxed{}}$$

Divide 18 and 30 by the common factor 2.

$$= \frac{9 \div \boxed{3}}{15 \div \boxed{3}}$$

Divide the numerator and denominator by the common factor 3.

$$= \frac{\boxed{}}{\boxed{}}$$

Since 3 and 5 have no common factors other than 1, stop dividing.

So, $\frac{18}{30}$ written in simplest form is $\dfrac{\boxed{}}{\boxed{}}$.

Guided Practice

Write each fraction in simplest form. If the fraction is already in simplest form, write *simplified*.

1. $\frac{4}{6}$

$$\frac{4}{6} = \frac{4 \div \boxed{2}}{6 \div \boxed{2}} = \frac{\boxed{}}{\boxed{}}$$

In simplest form, $\frac{4}{6} = \dfrac{\boxed{}}{\boxed{}}$.

2. $\frac{2}{12}$

$$\frac{2}{12} = \frac{2 \div \boxed{2}}{12 \div \boxed{2}} = \frac{\boxed{}}{\boxed{}}$$

In simplest form, $\frac{2}{12} = \dfrac{\boxed{}}{\boxed{}}$.

Talk MATH

Explain how to find the simplest form of any fraction.

Independent Practice

Write each fraction in simplest form. If the fraction is already in simplest form, write *simplified*.

3. $\frac{6}{8}$ _____

4. $\frac{6}{10}$ _____

5. $\frac{3}{18}$ _____

6. $\frac{2}{5}$ _____

7. $\frac{4}{16}$ _____

8. $\frac{12}{24}$ _____

9. $\frac{6}{25}$ _____

10. $\frac{21}{30}$ _____

11. $\frac{4}{11}$ _____

Algebra Find each unknown.

12. $\frac{8}{28} = \frac{m}{7}$

$m =$ _____

13. $\frac{12}{40} = \frac{b}{10}$

$b =$ _____

14. $\frac{9}{24} = \frac{3}{y}$

$y =$ _____

Problem Solving

15. PRACTICE  **Plan Your Solution** The table shows the results of a survey about favorite movie theater snacks. Write a fraction in simplest form that compares the number of people who chose popcorn to the total number of people surveyed.

Favorite Movie Snack	
Snack	Frequency
Popcorn	24
Hot dog	12
Nachos	11
Chocolate	8
Licorice	5

16. Kara buys 24 bagels. Ten are whole wheat. What fraction of the bagels are whole wheat? Write in simplest form.

HOT Problems

17. PRACTICE **Which One Doesn't Belong?** Circle the fraction that is not in simplest form. Explain.

$$\frac{9}{21} \qquad \frac{7}{18} \qquad \frac{3}{25} \qquad \frac{12}{31}$$

18. **Building on the Essential Question** Why is it important to be able to write a fraction in simplest form?

My Work!

Bagels put the hole in whole wheat.

MY Homework

Homework Helper

Need help? ⟋ connectED.mcgraw-hill.com

Amelia rode $\frac{8}{24}$ mile on the bike trail. Express in simplest form the fraction of the distance that Amelia rode on the bike trail.

Divide the numerator and denominator by the same common factor. Continue dividing until the fraction is in simplest form.

$$\frac{8}{24} = \frac{8 \div 2}{24 \div 2}$$ Divide 8 and 24 by the common factor 2.

$$= \frac{4}{12}$$ Simplify.

$$= \frac{4 \div 4}{12 \div 4}$$ Divide the numerator and denominator by the common factor 4.

$$= \frac{1}{3}$$ Simplify.

Since 1 and 3 have no common factors other than 1, stop dividing.

So, $\frac{8}{24}$ written in simplest form is $\frac{1}{3}$.

Practice

Write each fraction in simplest form. If the fraction is already in simplest form, write *simplified.*

1. $\frac{8}{9}$ _____

2. $\frac{9}{18}$ _____

3. $\frac{26}{32}$ _____

Problem Solving

Mathematical
4. **PRACTICE 3** **Find the Error** Nicholas wrote the steps below to simplify the fraction $\frac{20}{30}$. Find his error and correct it.

$$\frac{20}{30} = \frac{20 \div 5}{30 \div 6} = \frac{4}{5}$$

_____

_____

_____

_____

_____

5. Mr. Rolloson's class had a board game tournament. Out of the 24 games played, Timeka won 10 games. Write $\frac{10}{24}$ in simplest form.

_____

We are the winners!

Vocabulary Check

6. Fill in the blank with the correct term or number to complete the sentence.

A fraction is written in simplest form when the GCF of the numerator and the denominator is _____ .

Test Practice

7. Gil's aunt cut his birthday cake into 32 equal pieces. Eighteen pieces were eaten at his birthday party. What fraction of the cake was left?

Ⓐ $\frac{7}{16}$ Ⓒ $\frac{7}{12}$

Ⓑ $\frac{9}{16}$ Ⓓ $\frac{9}{14}$

Problem-Solving Investigation

STRATEGY: Guess, Check, and Revise

Lesson 4

ESSENTIAL QUESTION
How are factors and multiples helpful in solving problems?

Learn the Strategy

Watch | Tutor

The Bactrian camel has two humps, while the Dromedary camel has just one. Toby counted 20 camels with a total of 28 humps. How many camels of each type are there?

This is my best side!

1 Understand

What facts do you know?

- Bactrian camels have two humps and Dromedary camels have one hump.

- There are _____ camels with _____ humps.

What do you need to find?

- How many _____ of each type are there?

2 Plan

I will guess, check, and revise to solve the problem.

3 Solve

Bactrian Camels	Dromedary Camels	Number of Humps	Check
7	13	$(7 \times 2) + (13 \times 1) = 27$	too low
8	12	$(8 \times 2) + (12 \times 1) = 28$	correct

So, there are _____ Bactrian camels and _____ Dromedary camels.

4 Check

Is my answer reasonable?

_____ + _____ = 20 camels and _____ + _____ = 28 humps

Online Content at connectED.mcgraw-hill.com

Practice the Strategy

Conner spent $66 on rookie cards and Hall of Famer cards. How many of each type of card did he buy?

Baseball Card	Cost
Rookie	4 for $6
Hall of Famer	2 for $9

1 Understand

What facts do you know?

_____

_____

_____

What do you need to find?

_____

Let's trade!

2 Plan

_____

3 Solve

4 Check

Is my answer reasonable?

_____

_____

Name

Apply the Strategy

Guess, check, and revise to solve each problem.

1. Bike path A is 4 miles long. Bike path B is 7 miles long. If April biked a total of 37 miles, how many times did she bike each path?

2. Ruben sees 14 wheels on a total of 6 bicycles and tricycles. How many bicycles and tricycles are there?

3. A teacher is having three students take care of 28 goldfish during the summer. He gave some of them to Alaina. Then he gave twice as many to Miguel. He gave twice as many to Kira as he gave to Miguel. How many fish did each student get?

4. Ticket prices for a science museum are $18 for adults and $12 for students. If $162 is collected from a group of 12 people, how many adults and students are in the group?

5. Mathematical **PRACTICE** **1** **Keep Trying** Jerome bought 2 postcards and received $1.35 in change in quarters and dimes. If he got 6 coins back, how many of each coin did he get?

6. A tour director collected $258 for 13 tour packages. Tour package A costs $18 and tour package B costs $22. How many of each tour package were sold?

Review the Strategies

Use any strategy to solve each problem.
- Guess, check, and revise.
- Work backward.
- Determine an estimate or exact answer.
- Make a table.

7. The sum of two numbers is 30. Their product is 176. What are the two numbers?

_____

8. Mathematical **PRACTICE** 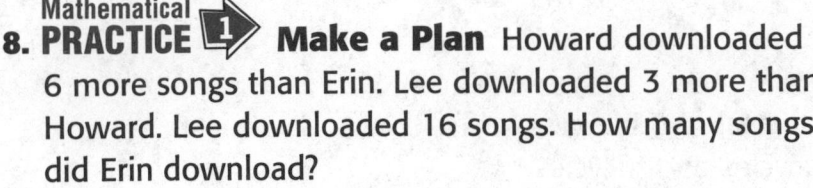 **Make a Plan** Howard downloaded 6 more songs than Erin. Lee downloaded 3 more than Howard. Lee downloaded 16 songs. How many songs did Erin download?

_____

9. Algebra Work backward to find the value of the variable in the equation below.

$$d - 4 = 23$$

_____

10. A bakery can make 175 pastries each day. The bakery has been sold out for 3 days in a row. Determine if an estimate or exact answer is needed. Then determine how many pastries were sold during the 3 days.

_____

11. Charlotte bought packages of hot dogs for $4 each. Each package contains 8 hot dogs. If she spent $16 on hot dogs, how many hot dogs did she buy?

_____

12. Mr. Thompson took his 5 children to the bowling alley. The cost for children 12 and older is $3.50. The cost for children under 12 is $2.25. He spends a total of $16.25. How many of his children are 12 and older?

_____

My Work!

Math is right up my ALLEY

MY Homework →

Homework Helper

Need help? connectED.mcgraw-hill.com

An automobile museum has cars and motorcycles displayed.
There are a total of 19 cars and motorcycles in the collection
and a total of 64 tires. How many cars and motorcycles are
in the collection?

1 Understand

What facts do you know?

• There are a total of 19 cars and motorcycles.

• There are a total of 64 tires.

What do you need to find?

• how many cars and motorcycles are in the collection

2 Plan

Guess, check, and revise to solve the problem.

3 Solve

Number of Cars	Number of Motorcycles	Number of Tires	Check
14	5	$(14 \times 4) + (5 \times 2) = 66$	too high
13	6	$(13 \times 4) + (6 \times 2) = 64$	correct

So, there are 13 cars and 6 motorcycles.

4 Check

Is my answer reasonable?

yes; $52 \div 4 = 13$, $12 \div 2 = 6$, and $13 + 6 = 19$.

Problem Solving

Guess, check, and revise to solve each problem.

1. A cabin has room for 7 campers and 2 counselors. How many cabins are needed for a total of 49 campers and 14 counselors?

2. Marcus is selling lemonade and peanuts. Each cup of lemonade costs $0.75, and each bag of peanuts costs $0.35. On Saturday, Marcus sold 5 more cups of lemonade than bags of peanuts. He earned $7.05. How many cups of lemonade did Marcus sell on Saturday?

3. **Mathematical PRACTICE 1 Keep Trying** In a recent year, a letter to Europe from the United States costs $0.94 to mail. A letter mailed within the United States costs $0.44. Nancy mailed 5 letters for $3.70, some to Europe and some to the United States. How many letters did she send to Europe?

4. Alma counts 26 legs in a barnyard with horses and chickens. If there are 8 animals, how many are horses?

Ready to practice?

5. A volleyball league organizer collected $2,040 for both divisions of volleyball teams. The Blue division costs $160 per team and the Red division costs $180 per team. How many teams will play in each division?

Check My Progress

Vocabulary Check

Draw lines to match each question with its correct answer.

1. Which is the greatest of the common factors of two or more numbers?

   • **fraction**

2. Which is a number that names equal parts of a whole or parts of a set?

   • **greatest common factor**

3. A fraction that is written when the greatest common factor of the numerator and denominator is 1 is written in which form?

   • **simplest form**

Concept Check

Represent each situation using a model. Then solve.

4. Three gallons of paint are used to paint 16 wooden signs. How much paint did each sign use? Between what two whole numbers does the answer lie?

   _____

   _____

5. Refer to Exercise 4. How many wooden signs can be painted with one gallon of paint? Between what two whole numbers does the answer lie?

   _____

   _____

Find the GCF of each set of numbers.

6. 36, 90 _____

7. 16, 24, 40 _____

Write each fraction in simplest form. If the fraction is already in simplest form, write *simplified*.

8. $\frac{4}{14}$ _____

9. $\frac{15}{20}$ _____

10. $\frac{21}{35}$ _____

Problem Solving

11. Anja buys a magazine and a pizza. She spends $9.75. The pizza costs twice as much as the magazine. How much does the pizza cost? Guess, check, and revise to solve.

_____

12. In a pile of coins, there are 7 more quarters than nickels. If there is a total of $2.65 in coins, how many quarters are there? Guess, check, and revise to solve.

_____

Hot off the shelf!

Test Practice

13. A warehouse has shelves that can hold 8, 12, or 16 skateboards. Each shelf has sections holding the same number of skateboards. What is the greatest number of skateboards that can be put in a section?

Ⓐ 1 skateboard   Ⓒ 3 skateboards

Ⓑ 2 skateboards   Ⓓ 4 skateboards

Number and Operations – Fractions
Preparation for 5.NF.2

Least Common Multiple

Lesson 5

ESSENTIAL QUESTION
How are factors and multiples helpful in solving problems?

A **multiple** of a number is the product of the number and any other whole number (0, 1, 2, 3, 4, . . .). Multiples that are shared by two or more numbers are **common multiples**.

The **least common multiple (LCM)** is the least multiple, other than 0, common to sets of multiples.

Math in My World

Example 1

Tom hits golf balls at the driving range every 3 days, practices putting every 4 days, and goes golfing every 6 days. If he did all three activities today, in how many days will he do all three activities again?

Find and circle the LCM of 3, 4 and 6 by listing nonzero multiples of each number.

3: _____

4: _____

6: _____

So, Tom will complete all three activities again in _____ days.

Check The number line shows on which days all three activities will be completed.

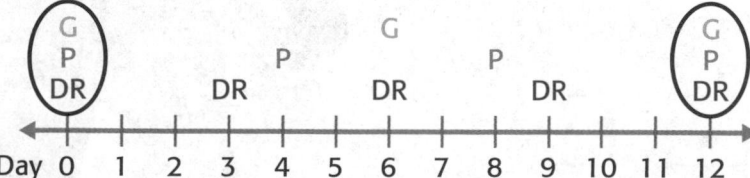

G = golf
P = putting
DR = driving range

Example 2
Find the LCM of 15 and 40.

 Write the prime factorization of each number.

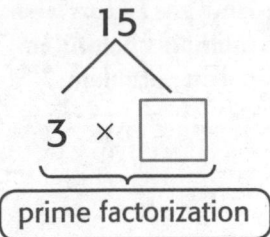

15

3 × ☐

prime factorization

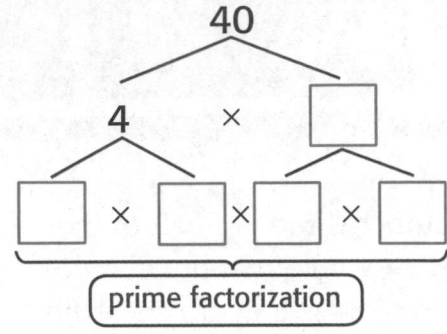

40

4 × ☐

☐ × ☐ | ☐ × ☐

prime factorization

② Find the common prime factors.

15 = ____ × ____

40 = ____ × ____ × ____ × ____

← The only common prime factor is _____.

③ Find the product of the prime factors using each common prime factor only once and any remaining factors.

The LCM is _____ × _____ × _____ × _____ × _____, or _____.

Guided Practice

Find the LCM of each set of numbers.

1. 6, 10

6: _____

10: _____

So, the LCM of 6 and 10 is _____.

2. 3, 4

3: _____

4: _____

So, the LCM of 3 and 4 is _____.

Talk MATH
Could the LCM of two numbers be one of the numbers? Explain.

Name ....

Independent Practice

Find the LCM of each set of numbers.

3. 2, 13 _____   **4.** 7, 9 _____   **5.** 2, 10 _____

6. 12, 15 _____   **7.** 16, 20 _____   **8.** 3, 8 _____

9. 4, 8, 10 _____   **10.** 3, 9, 18 _____   **11.** 15, 25, 75 _____

12. 9, 12, 15 _____   **13.** 4, 7, 10 _____   **14.** 6, 7, 9 _____

Problem Solving

15. Juan goes to the bowling alley every 3 weeks. Percy goes to the bowling alley every 5 weeks. If Juan and Percy meet the first time they go bowling, how many weeks will it be before they see each other at the bowling alley again?

_____

16. A full moon occurs about every 30 days. If the last full moon occurred on a Friday, how many days will pass before a full moon occurs again on a Friday?

_____

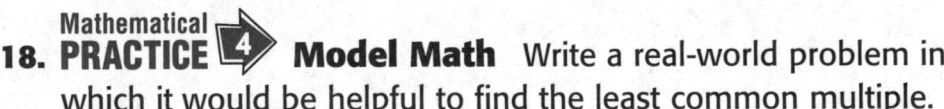

HOT Problems

17. **Mathematical PRACTICE 3** **Find the Error** Maria is finding the LCM of 6 and 8. Help find and correct her mistake.

$$6 = \boxed{2} \times 3$$
$$8 = \boxed{2} \times 2 \times 2$$

The LCM of 6 and 8 is 2.

18. **Mathematical PRACTICE 4** **Model Math** Write a real-world problem in which it would be helpful to find the least common multiple.

_____

_____

_____

19. **Building on the Essential Question** How can I find the least common multiple of two or more numbers?

_____

_____

_____

MY Homework

Homework Helper

Need help? connectED.mcgraw-hill.com

Ben's Burgers gives away a free order of fries every 2 days, a free milkshake every 3 days, and a free hamburger every 4 days. If they gave away all three items today, in how many days will they give away all three items again?

Find and circle the LCM of 2, 3 and 4 by listing nonzero multiples of each number.

2: 2, 4, 6, 8, 10, (12) . . .
3: 3, 6, 9, (12), 15 . . .
4: 4, 8, (12), 16, 20 . . .

The least common multiple of 2, 3, and 4 is 12.

So, Ben's Burgers will give away all three items again in 12 days.

Check The number line shows on which days all three activities will be completed.

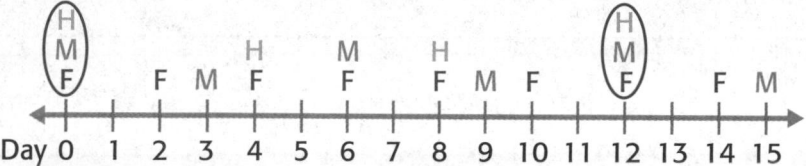

H = hamburger
M = milkshake
F = fries

Practice

Find the LCM of each set of numbers.

1. 7, 14 _____

2. 6, 15 _____

3. 5, 9, 15 _____

Problem Solving

4. The cycles for two different events are shown in the table. Each of these events happened in the year 2000. What is the next year in which both will happen?

Event	Cycle (yr)
Summer Olympics	4
United States Census	10

_____

5. Mathematical PRACTICE 3 Draw a Conclusion Is the statement below _always, sometimes,_ or _never_ true? Give at least two examples to support your reasoning. The LCM of two numbers is the product of the two numbers.

_____

_____

Vocabulary Check

Fill in each blank with the correct word(s) to complete each sentence.

6. Multiples that are shared by two or more numbers are

_____.

7. The least common multiple (LCM) is the _____ multiple, other than 0, common to sets of multiples.

Test Practice

8. Micah is buying items for a birthday party. If he wants to have the same amount of each item, what is the least number of packages of cups he needs to buy?

Party Supplies	
Item	**Number in Each Package**
Cups	6
Plates	8

(A) 2 packages   (C) 4 packages

(B) 3 packages   (D) 5 packages

Name _____

Compare Fractions

Lesson 6

ESSENTIAL QUESTION
How are factors and multiples helpful in solving problems?

The **least common denominator (LCD)** is the LCM of the denominators of the fractions. You can use the LCD to compare fractions.

Math in My World
 Tools Watch Tutor

I get a kick out of math!

Example 1

Trevor made 2 out of 3 goals and Tyler made 5 out of 6. Who made a greater fraction of goals?

1 Find the LCM of the denominators of $\frac{2}{3}$ and $\frac{5}{6}$.

3: _____

6: _____ ⟵ | The LCM of 3 and 6 is _____. |

2 Find equivalent fractions with a denominator of _____.

$$\frac{2}{3} = \frac{2 \times \boxed{2}}{3 \times \boxed{2}} = \frac{\boxed{}}{\boxed{}} \qquad \frac{5}{6} = \frac{5 \times \boxed{1}}{6 \times \boxed{1}} = \frac{\boxed{}}{\boxed{}}$$

3 Compare the numerators.

Since 4 < 5, $\frac{\boxed{}}{\boxed{}} < \frac{\boxed{}}{\boxed{}}$. So, $\frac{2}{3} \bigcirc \frac{5}{6}$.

Helpful Hint
Multiplying the numerator and denominator by the same number is the same as multiplying the fraction by 1. The result is an equivalent fraction.

So, _____ made a greater fraction of goals.

Check

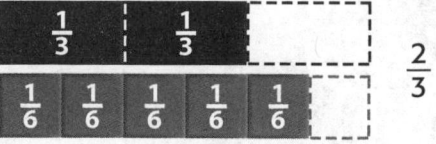

$\frac{2}{3} < \frac{5}{6}$

Online Content at **connectED.mcgraw-hill.com**

Example 2

Compare $\frac{3}{5}$ and $\frac{1}{2}$ using the least common denominator.

 Find the LCM of the denominators.

5: _____

2: _____ ← The LCM of 2 and 5 is _____ .

 Find equivalent fractions with a denominator of _____ .

$\frac{3}{5} = \frac{3 \times 2}{5 \times 2} = \frac{\boxed{}}{\boxed{}}$ $\frac{1}{2} = \frac{1 \times 5}{2 \times 5} = \frac{\boxed{}}{\boxed{}}$

 Compare the numerators.

Since 6 > 5, $\dfrac{\boxed{}}{\boxed{}} > \dfrac{\boxed{}}{\boxed{}}$. So, $\frac{3}{5} \bigcirc \frac{1}{2}$.

Check The models show that $\frac{3}{5} > \frac{1}{2}$.

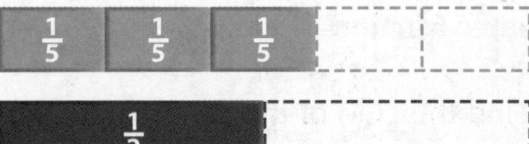

Guided Practice

1. Compare $\frac{1}{5}$ and $\frac{1}{3}$ using the LCD.

5: _____

3: _____

$\frac{1}{5} = \frac{1 \times 3}{5 \times 3} = \frac{\boxed{}}{\boxed{}}$

$\frac{1}{3} = \frac{1 \times 5}{3 \times 5} = \frac{\boxed{}}{\boxed{}}$

So, $\frac{1}{5} \bigcirc \frac{1}{3}$.

Talk MATH

Explain how the LCM and the LCD are alike. How are they different?

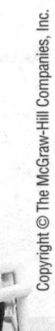

Name _____

Independent Practice

Mathematical PRACTICE **6** **Be Precise** Compare each pair of fractions by drawing models or using the LCD. Use the symbols $<$, $>$, or $=$.

2. $\frac{3}{4}$ ◯ $\frac{7}{8}$

3. $\frac{2}{3}$ ◯ $\frac{7}{10}$

4. $\frac{2}{3}$ ◯ $\frac{7}{12}$

5. $\frac{1}{3}$ ◯ $\frac{5}{9}$

6. $\frac{1}{4}$ ◯ $\frac{1}{6}$

7. $\frac{2}{5}$ ◯ $\frac{6}{15}$

8. $\frac{2}{3}$ ◯ $\frac{3}{4}$

9. $\frac{1}{5}$ ◯ $\frac{3}{15}$

10. $\frac{1}{6}$ ◯ $\frac{1}{3}$

Algebra Find each unknown in each equation that shows equivalent fractions.

11. $\dfrac{3 \times m}{4 \times 5} = \dfrac{p}{20}$

$m =$ _____

$p =$ _____

12. $\dfrac{7 \times g}{8 \times k} = \dfrac{21}{24}$

$g =$ _____

$k =$ _____

13. $\dfrac{5}{6} = \dfrac{b}{48}$

$b =$ _____

Problem Solving

14. The amounts of water four runners drank are shown at the right. Who drank the most?

Runner	Amount (bottle)
Evita	$\frac{3}{5}$
Jack	$\frac{5}{8}$
Keisha	$\frac{3}{4}$
Sirjo	$\frac{5}{10}$

_____

Mathematical
15. PRACTICE 6 **Be Precise** A recipe calls for $\frac{5}{8}$ cup of brown sugar and $\frac{2}{3}$ cup of flour. Which ingredient has the greater amount?

_____

16. A trail mix has $\frac{1}{2}$ cup of raisins and $\frac{2}{3}$ cup of peanuts. Which ingredient has the greater amount?

_____

Let's hit the trail!

Mathematical
17. PRACTICE 2 **Use Number Sense** Explain why multiplying the numerator and denominator of a fraction by the same number results in an equivalent fraction.

_____

_____

_____

18. ? **Building on the Essential Question** What is one way to compare fractions with unlike denominators?

_____

_____

_____

MY Homework

Homework Helper

Need help? connectED.mcgraw-hill.com

Compare $\frac{4}{5}$ and $\frac{5}{6}$ using the least common denominator.

 Find the LCM of the denominators.

5: 5, 10, 15, 20, 25, 30 . . .

6: 6, 12, 18, 24, 30 . . . ← The LCM of 5 and 6 is 30.

 Find equivalent fractions with a denominator of 30.

$$\frac{4}{5} = \frac{4 \times 6}{5 \times 6} = \frac{24}{30} \qquad \frac{5}{6} = \frac{5 \times 5}{6 \times 5} = \frac{25}{30}$$

 Compare the numerators.

Since 24 < 25, then $\frac{24}{30} < \frac{25}{30}$. So, $\frac{4}{5} < \frac{5}{6}$.

Check The models show that $\frac{4}{5} < \frac{5}{6}$.

Practice

Compare each pair of fractions by drawing models or using the LCD. Use the symbols <, >, or =.

1. $\frac{3}{4}$ ◯ $\frac{7}{8}$

2. $\frac{1}{3}$ ◯ $\frac{3}{9}$

3. $\frac{3}{4}$ ◯ $\frac{2}{3}$

Problem Solving

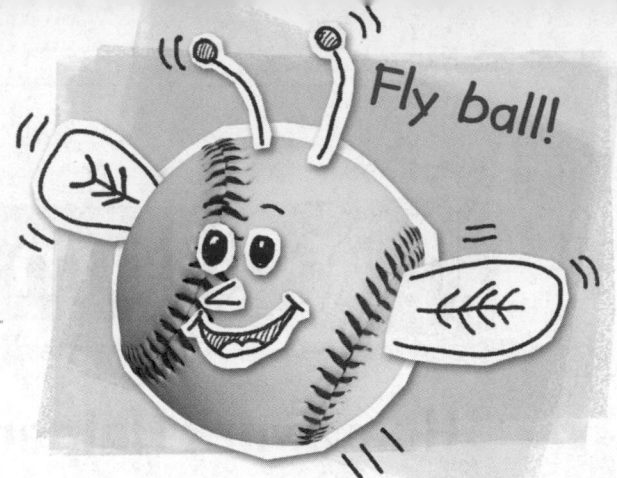

Fly ball!

4. A survey showed that $\frac{7}{15}$ of a class liked soccer and $\frac{2}{5}$ liked baseball. Which sport was liked less?

_____

5. The fifth graders were given sandwiches for lunch during their field trip. Nathan ate $\frac{5}{6}$ of his sandwich, Leroy ate $\frac{7}{8}$ of his sandwich, and Sofia ate $\frac{5}{8}$ of her sandwich. Who ate the greatest amount of their sandwich?

_____

6. Mathematical **PRACTICE**  **Use Symbols** Replace ■ with a number to make $\frac{\blacksquare}{24} > \frac{1}{4}$ a true statement.

_____

Vocabulary Check

7. Fill in the blank with the correct word to complete the sentence. The least common denominator (LCD) is the least common

multiple of the _____ of the fractions.

Test Practice

8. Eighteen out of 24 of Emil's CDs are country music. Five out of 8 of Imani's CDs are country music. Which is a true statement?

Ⓐ Half of each CD collection consists of country music.

Ⓑ Less than half of each CD collection consists of country music.

Ⓒ Emil has a greater fraction of country music than Imani.

Ⓓ Imani has a greater fraction of country music than Emil.

Hands On
Use Models to Write Fractions as Decimals

Lesson 7
ESSENTIAL QUESTION
How are factors and multiples helpful in solving problems?

You can use models to write fractions as equivalent decimals.

Draw It

Use a model to write $\frac{1}{2}$ as a decimal.

 Write $\frac{1}{2}$ as an equivalent fraction with a denominator of 10.

$$\frac{1}{2} = \frac{1 \times \boxed{5}}{2 \times \boxed{5}} = \frac{\square}{\square}$$

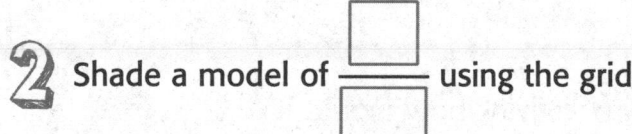 Shade a model of $\frac{\square}{\square}$ using the grid.

How many tenths are shaded? _____

The model shows _____ *tenths* or _____.

So, $\frac{1}{2} =$ _____ .

Helpful Hint
Multiplying $\frac{1}{2}$ by $\frac{5}{5}$ is the same as multiplying $\frac{1}{2}$ by 1. The result is an equivalent fraction.

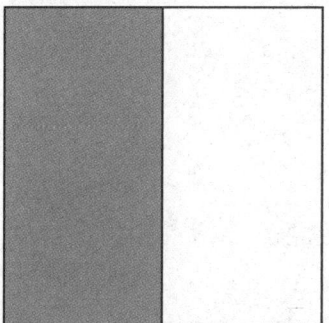

Try It

Use a model to write $\frac{3}{4}$ as a decimal.

 Write $\frac{3}{4}$ as a fraction with a denominator of 100.

$$\frac{3}{4} = \frac{3 \times \boxed{25}}{4 \times \boxed{25}} = \frac{\boxed{}}{\boxed{}}$$

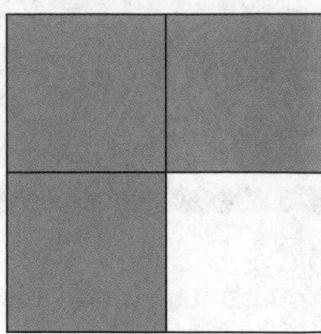

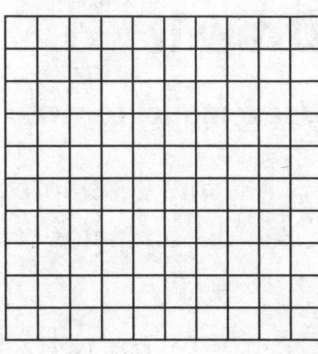

② Shade a model of $\dfrac{\boxed{}}{\boxed{}}$ using the 10-by-10 grid.

How many squares out of the 100 are shaded? _____

The model shows _____ *hundredths* or _____ .

So, $\frac{3}{4} =$ _____ .

Talk About It

1. **Mathematical PRACTICE** 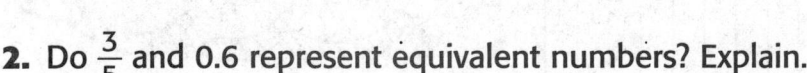 **Model Math** In the first activity, how would it change if $\frac{1}{2}$ was written as a fraction with a denominator of 100? Would the result be the same? Explain.

_____

_____

_____

2. Do $\frac{3}{5}$ and 0.6 represent equivalent numbers? Explain.

_____

_____

Practice It

Mathematical
PRACTICE 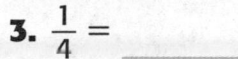 Use Math Tools Shade each model.
Then write each fraction as a decimal.

3. $\frac{1}{4} =$ _____

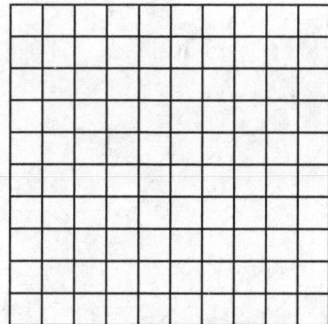

4. $\frac{3}{20} =$ _____

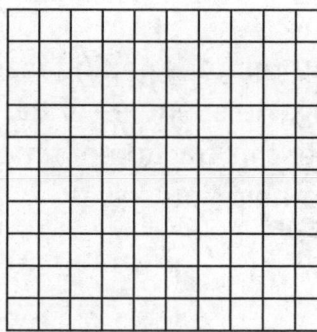

5. $\frac{2}{5} =$ _____

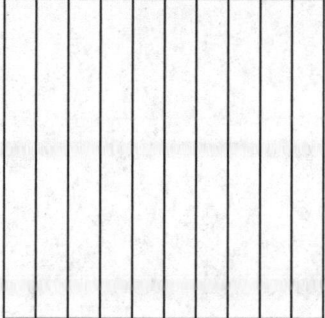

6. $\frac{3}{5} =$ _____

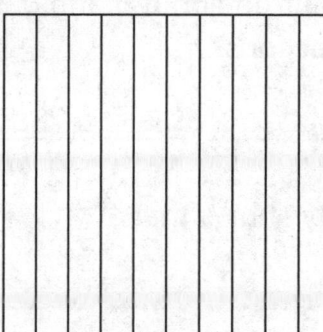

7. $\frac{7}{10} =$ _____

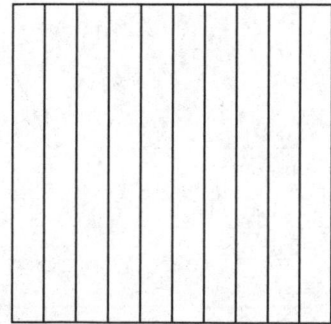

8. $\frac{8}{25} =$ _____

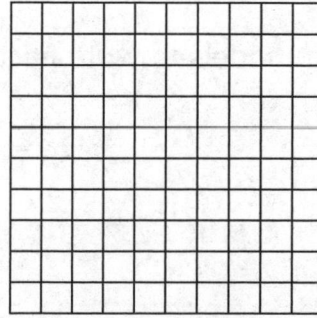

Apply It

9. Juanita practiced shooting 25 free throws at basketball practice. She made $\frac{17}{25}$ of the attempts. Write the fraction of attempts made as a decimal. Use models to help you solve.

_____

My Work!

10. Travis spent 20 minutes getting ready for school in the morning. He spent $\frac{9}{20}$ of the time eating breakfast. Write this fraction of time as a decimal. Use models to help you solve.

_____

Mathematical PRACTICE 2 Use Algebra For Exercises 11–13, refer to the equation $\frac{2 \times p}{5 \times q} = \frac{40}{100}$.

11. What must be true about p and q if the equation shows equivalent fractions?

_____

12. What property shows that $\frac{2}{5} \times 1 = \frac{40}{100}$?

_____

13. Write the decimal equivalent for $\frac{2}{5}$ and $\frac{40}{100}$.

_____

Write About It

14. How can I use models to write fractions as decimals?

_____

_____

_____

Name ....

MY Homework

Homework Helper

Need help? connectED.mcgraw-hill.com

Use a model to write $\frac{7}{20}$ as a decimal.

1 Write $\frac{7}{20}$ as a fraction with a denominator of 100.

$$\frac{7}{20} = \frac{7 \times 5}{20 \times 5} = \frac{35}{100}$$

Helpful Hint
Multiplying the numerator and denominator by the same number is the same as multiplying the fraction by 1. The result is an equivalent fraction.

2 The model represents $\frac{35}{100}$.

There are 35 squares out of the 100 that are shaded.

The model shows 35 *hundredths* or 0.35.

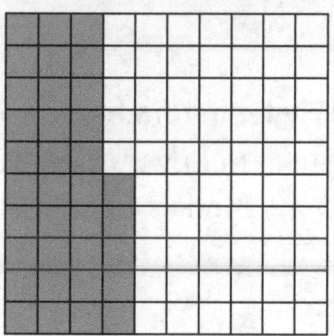

So, $\frac{7}{20} = 0.35$.

Practice

Shade each model. Then write each fraction as a decimal.

1. $\frac{9}{10} =$ _____

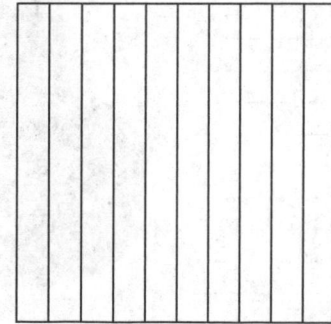

2. $\frac{11}{20} =$ _____

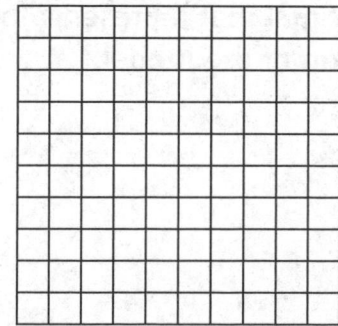

Problem Solving

3. Terrell hit a total of 10 home runs during the baseball season. He hit $\frac{4}{5}$ of the home runs during the first half of the season. Write the fraction of home runs hit during the first half of the season as a decimal. Draw models to help you solve.

_____

4. **Mathematical PRACTICE 5 Use Math Tools** Bradley and his family drove to visit a museum. They drove $\frac{9}{25}$ of the way and stopped to get gasoline. Write the fraction of the distance traveled as a decimal. Draw models to help you solve.

_____

5. Lilly let her friend borrow $\frac{1}{10}$ of the money in her purse to buy a snack. Write the fraction as a decimal. Draw models to help you solve.

_____

6. Jackson was playing chess. Out of all the games he played, he won $\frac{7}{25}$ of the time. Write this fraction as a decimal. Draw models to help you solve.

_____

Queen for the day!

7. Write the decimal that represents the shaded portion of the model.

_____

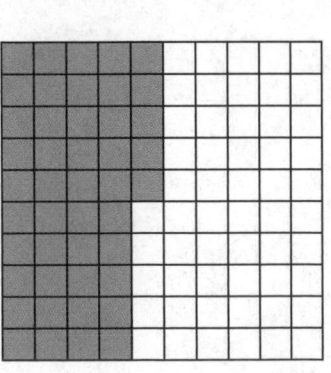

Write Fractions as Decimals

Lesson 8

ESSENTIAL QUESTION
How are factors and multiples helpful in solving problems?

 ## Math in My World
Watch Tutor

Example 1

The average weight of a tennis racquet is $\frac{2}{5}$ pound. Write this weight as a decimal.

Write $\frac{2}{5}$ as a decimal.

1 Write $\frac{2}{5}$ as an equivalent fraction with a denominator of 10.

$$\frac{2}{5} = \frac{2 \times \boxed{2}}{5 \times \boxed{2}} = \frac{\boxed{}}{\boxed{}}$$ Since 5 × 2 = 10, multiply 2 × 2 to obtain 4.

2 Write the fraction with a denominator of 10 as a decimal.

$$\frac{2}{5} = \underline{\hspace{2cm}}$$ $\frac{4}{10}$ means *four tenths,* or 0.4.

You can use a place-value chart to read the decimal.

Read the decimal. It reads four _____.

The average weight of a tennis racquet is _____ pound.

Ones	Tenths	Hundredths
0	4	

Example 2

Write $\frac{3}{4}$ as a decimal.

 Write $\frac{3}{4}$ as an equivalent fraction with a denominator of 100.

Helpful Hint
Multiplying the numerator and denominator by the same number is the same as multiplying the fraction by 1. The result is an equivalent fraction.

$$\frac{3}{4} = \frac{3 \times \boxed{25}}{4 \times \boxed{25}} = \frac{\boxed{}}{\boxed{}}$$ Since $4 \times 25 = 100$, multiply 3×25 to obtain 75.

 Write the fraction with a denominator of 100 as a decimal.

$$\frac{3}{4} = \underline{\hspace{2cm}}$$ $\frac{75}{100}$ means *seventy-five hundredths,* or 0.75.

Read the decimal as _____ .

Guided Practice ✓Check

1. Write $\frac{1}{5}$ as a decimal.

$$\frac{1}{5} = \frac{1 \times \boxed{2}}{5 \times \boxed{2}} = \frac{\boxed{}}{\boxed{}}$$

So, $\frac{1}{5} =$ _____

Read the decimal as _____ .

Talk MATH

Explain how to write a fraction as a decimal using equivalent fractions.

2. Write $\frac{11}{25}$ as a decimal.

$$\frac{11}{25} = \frac{11 \times \boxed{4}}{25 \times \boxed{4}} = \frac{\boxed{}}{\boxed{}}$$

So, $\frac{11}{25} =$ _____

Read the decimal as _____ .

Independent Practice

Write each fraction as a decimal.

3. $\frac{8}{10} =$ _.8_

4. $\frac{1}{20} = \frac{5}{100} = .05$

5. $\frac{17}{20} = \frac{85}{100} .85$

6. $\frac{4}{25} = \frac{16}{100} .16$

7. $\frac{1}{10} =$ _.1_

8. $\frac{8}{25} =$ _____

9. $\frac{14}{25} =$ _____

10. $\frac{1}{4} =$ _____

11. $\frac{7}{20} =$ _____

12. $\frac{1}{25} =$ _____

13. $\frac{9}{10} =$ _____

14. $\frac{9}{25} =$ _____

Algebra Find each unknown.

15. $\frac{g}{20} = 0.65$

$g =$ _____

16. $0.7 = \frac{7}{w}$

$w =$ _____

17. $\frac{n}{50} = 0.18$

$n =$ _____

Problem Solving

18. The smallest known female spider is $\frac{23}{50}$ millimeter long. The smallest male spider is $\frac{37}{100}$ millimeter long. Write each fraction as a decimal.

_____

19. **PRACTICE 4 Model Math** Evan drank $\frac{2}{25}$ gallon of water throughout the day. Write $\frac{2}{25}$ as a decimal.

_____

20. At hockey practice, Savannah spent $\frac{19}{20}$ hour practicing passing. Write $\frac{19}{20}$ as a decimal.

_____

HOT Problems

21. Mathematical **PRACTICE 3 Find the Error** Juliana wrote the steps below to write the fraction $\frac{18}{25}$ as a decimal. Find her error and correct it.

$$\frac{18}{25} = \frac{18 \times 2}{25 \times 4} = \frac{36}{100} = 0.36$$

_____

_____

_____

_____

22. **Building on the Essential Question** What is the relationship between fractions and decimals?

_____

_____

_____

MY Homework

Lesson 8

Write Fractions as Decimals

Homework Helper

Need help? connectED.mcgraw-hill.com

The average length of a honeybee is $\frac{4}{5}$ inch. Write $\frac{4}{5}$ as a decimal.

1 Write $\frac{4}{5}$ as an equivalent fraction with a denominator of 10.

$$\frac{4}{5} = \frac{4 \times \boxed{2}}{5 \times \boxed{2}} = \frac{8}{10}$$ Since $5 \times 2 = 10$, multiply 4×2 to obtain 8.

2 Write the fraction with a denominator of 10 as a decimal.

$$\frac{4}{5} = 0.8$$

Read the decimal as *eight tenths*.

Practice

Write each fraction as a decimal.

1. $\frac{1}{2} =$ _____

2. $\frac{11}{20} =$ _____

3. $\frac{13}{20} =$ _____

4. $\frac{6}{10} =$ _____

5. $\frac{13}{25} =$ _____

6. $\frac{14}{20} =$ _____

Problem Solving

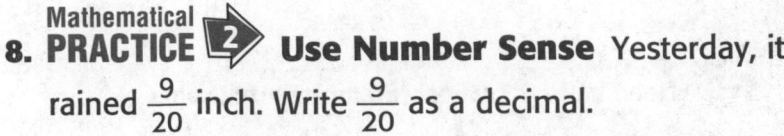

7. Courtney hit the bullseye $\frac{3}{5}$ of the time when playing darts. Write $\frac{3}{5}$ as a decimal.

_____

8. **Mathematical PRACTICE**  **Use Number Sense** Yesterday, it rained $\frac{9}{20}$ inch. Write $\frac{9}{20}$ as a decimal.

_____

9. Camille shaded $\frac{12}{25}$ of a model. Write the decimal that represents the shaded portion of the model.

_____

10. Paolo built a model car that is $\frac{7}{25}$ the size of his dad's car. Write $\frac{7}{25}$ as a decimal.

_____

11. Bishon sold $\frac{3}{20}$ of his collection of sports cards. Write $\frac{3}{20}$ as a decimal.

_____

Test Practice

12. Emilia bought $\frac{9}{12}$ pound of sliced salami at the deli counter. Which of the following decimals did the scale show?

Ⓐ 0.25 pound

Ⓑ 0.34 pound

Ⓒ 0.7 pound

Ⓓ 0.75 pound

Review

Vocabulary Check

Use the word bank below to complete each sentence.

common factor	common multiple
denominator	equivalent fractions
fraction	greatest common factor (GCF)
least common multiple (LCM)	least common denominator (LCD)
multiple	numerator
simplest form	

1. The _____ is the least multiple, other than 0, common to sets of multiples.

2. Fractions that name the same number are _____.

3. The top number of a fraction is called the _____.

4. When the numerator and the denominator have no common factor greater than 1, the fraction is written in _____.

5. The bottom number of a fraction is called the _____.

6. A whole number that is a factor of two or more numbers is called a(n) _____.

7. The greatest of the common factors of two or more numbers is the _____ of the numbers.

8. The _____ is the least common multiple of the denominators of the fractions.

Concept Check ✓

Find the GCF of each set of numbers.

9. 11, 44 _____

10. 12, 21, 30 _____

Write each fraction in simplest form. If the fraction is already in simplest form, write *simplified*.

11. $\frac{3}{36}$ _____

12. $\frac{25}{30}$ _____

Find the LCM of each set of numbers.

13. 4, 9 _____

14. 5, 7, 10 _____

Compare each pair of fractions using models or the LCD. Use the symbols <, >, or =.

15. $\frac{2}{5}$ ◯ $\frac{3}{10}$

16. $\frac{1}{5}$ ◯ $\frac{1}{4}$

17. $\frac{3}{4}$ ◯ $\frac{18}{24}$

Write each fraction as a decimal.

18. $\frac{3}{10}$ = _____

19. $\frac{19}{50}$ = _____

20. $\frac{5}{10}$ = _____

21. $\frac{1}{5}$ = _____

22. $\frac{14}{25}$ = _____

23. $\frac{3}{25}$ = _____

Problem Solving

24. Three bags of packing peanuts are used to fill 2 boxes. How many bags of packing peanuts does each box use? Between which two whole numbers does the answer lie?

_____

_____

_____

25. A van has room for 6 students and 2 teachers. How many vans are needed for a total of 48 students and 16 teachers?

_____

26. The table shows the number of each type of toy in a store. The toys will be placed on shelves so that each shelf has the same number of each type of toy. How many shelves are needed for each type of toy so that it has the greatest number of toys?

Toy	Amount
Dolls	45
Footballs	105
Small cars	75

_____

_____

Test Practice

27. In a typical symphony orchestra, 16 out of every 100 musicians are violinists. What fraction of the orchestra are violinists?

Ⓐ $\frac{2}{25}$ Ⓒ $\frac{1}{5}$

Ⓑ $\frac{4}{25}$ Ⓓ $\frac{8}{25}$

Reflect

Use what you learned about fractions and decimals to complete the graphic organizer.

Write an Example

Real-World Example

ESSENTIAL QUESTION

How are factors and multiples helpful in solving problems?

Vocabulary

Prime Factorization

Now reflect on the ESSENTIAL QUESTION Write your answer below.

Chapter 9 — Add and Subtract Fractions

ESSENTIAL QUESTION

How can equivalent fractions help me add and subtract fractions?

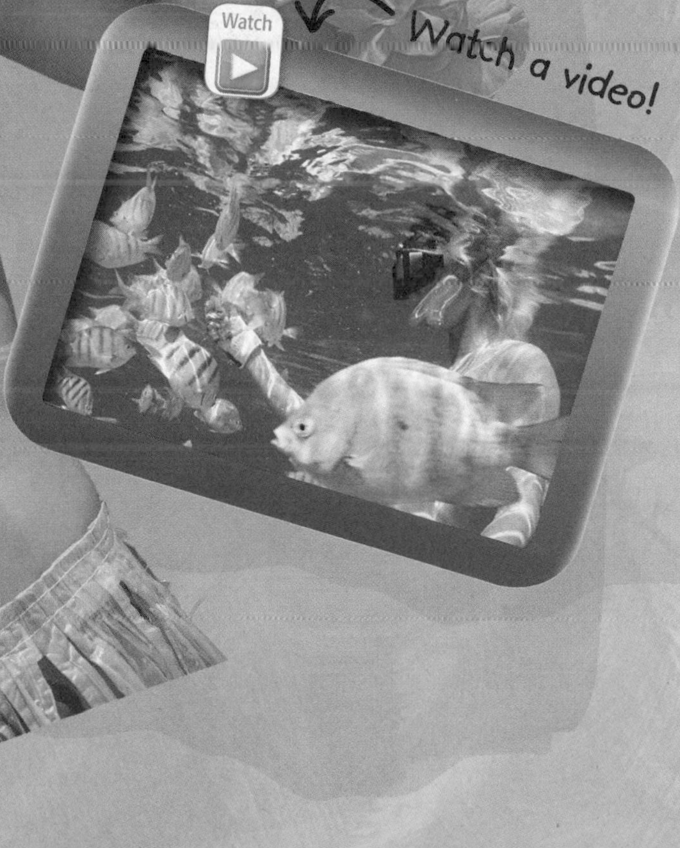

Our Oceans

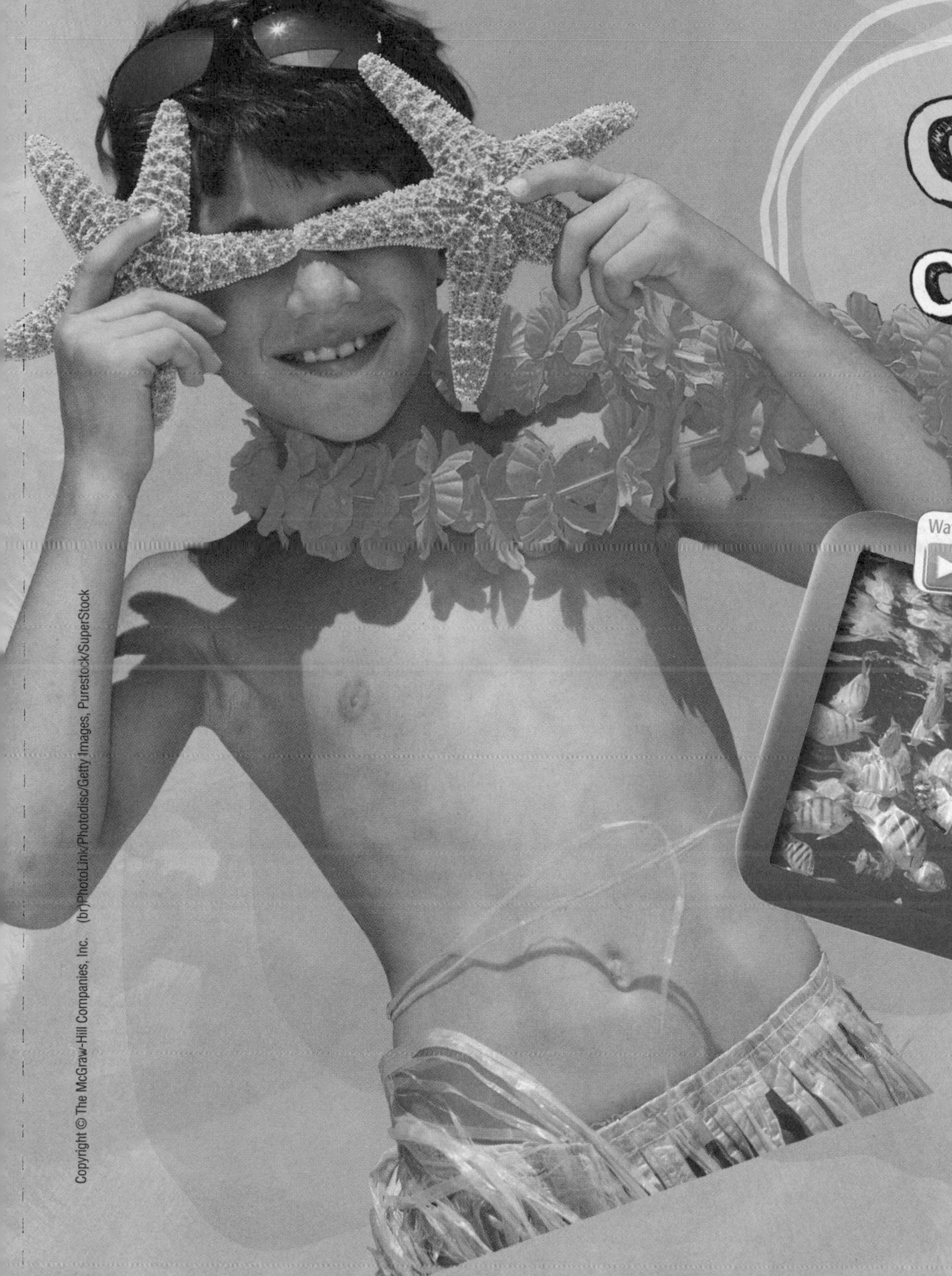

Watch a video!

MY Standards

Number and Operations – Fractions

5.NF.1 Add and subtract fractions with unlike denominators (including mixed numbers) by replacing given fractions with equivalent fractions in such a way as to produce an equivalent sum or difference of fractions with like denominators.

5.NF.2 Solve word problems involving addition and subtraction of fractions referring to the same whole, including cases of unlike denominators, e.g., by using visual fraction models or equations to represent the problem. Use benchmark fractions and number sense of fractions to estimate mentally and assess the reasonableness of answers.

Hmm, I don't think this will be too bad!

Standards for Mathematical PRACTICE

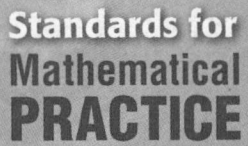

1. Make sense of problems and persevere in solving them.
2. Reason abstractly and quantitatively.
3. Construct viable arguments and critique the reasoning of others.
4. Model with mathematics.
5. Use appropriate tools strategically.
6. Attend to precision.
7. Look for and make use of structure.
8. Look for and express regularity in repeated reasoning.

= focused on in this chapter

Name

Am I Ready?

Check ✓ ← Go online to take the Readiness Quiz

Write each fraction in simplest form.

1. $\frac{4}{8}$ = _____

2. $\frac{4}{12}$ = _____

3. $\frac{15}{20}$ = _____

4. $\frac{4}{24}$ = _____

5. $\frac{16}{30}$ = _____

6. $\frac{24}{40}$ = _____

7. Marcela made 4 out of 16 free throws. Write the fraction of free throws she made in simplest form.

_____

Write each improper fraction as a mixed number.

8. $\frac{10}{7}$ = _____

9. $\frac{3}{2}$ = _____

10. $\frac{14}{6}$ = _____

11. $\frac{22}{4}$ = _____

12. $\frac{30}{7}$ = _____

13. $\frac{41}{8}$ = _____

14. A recipe for potato casserole calls for $\frac{7}{4}$ cups of cheese. Write the fraction as a mixed number.

_____

Shade the boxes to show the problems you answered correctly.

How Did I Do?	1	2	3	4	5	6	7	8	9	10	11	12	13	14

MY Math Words

Vocab
abc

Review Vocabulary

factors   greatest common factor (GCF)
least common multiple (LCM)   mixed numbers   multiples

Making Connections

Use the review words to complete the diagram below which
relates factors and multiples of the numbers 9 and 12.

Factors and Multiples

_____ of 9 and 12

9: 1, 3, 9

12: 1, 2, 3, 4, 6, 12

What do the circled
numbers represent?

_____

_____ of 9 and 12

9: 0, 9, 18, 27, 36

12: 0, 12, 24, 36, 48

What do the circled
numbers represent?

_____

How can greatest common factors and least common multiples
help you work with fractions?

_____

_____

MY Vocabulary Cards

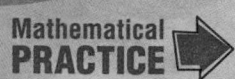

like fractions

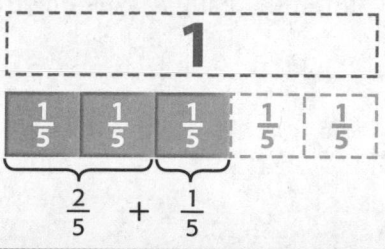

unlike fractions

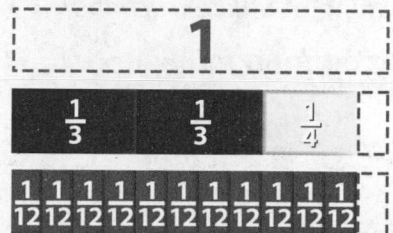

Ideas for Use

- Practice your penmanship! Write this chapter's essential question and each word in cursive.

- Use the back of the card to write or draw examples to help you answer the essential question.

Fractions that have different denominators.

How can the prefix *un-* help you remember the meaning of *unlike fractions*?

_____

_____

_____

Fractions that have the same denominator.

Like is used as an adjective in *like fractions*. Write a sentence using *like* as a verb.

_____

_____

MY Foldable

FOLDABLES® Follow the steps on the back to make your Foldable.

Like Denominators

$$\frac{3}{8} + \frac{1}{8}$$

1

2

3

Unlike Denominators

$$\frac{3}{10} + \frac{1}{5}$$

1

2

3

 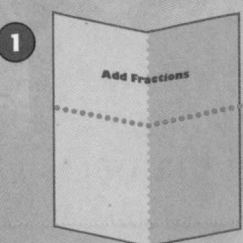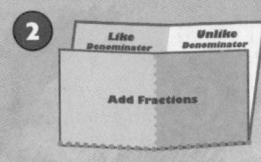

Round Fractions

Lesson 1

ESSENTIAL QUESTION
How can equivalent fractions help me add and subtract fractions?

You can use benchmark fractions, such as $\frac{1}{2}$, to round a fraction to 0, $\frac{1}{2}$, or 1.

Math in My World

 Watch Tools Tutor

Example 1

A poison dart frog is 2 inches long. This is equal to $\frac{2}{12}$ foot. Is $\frac{2}{12}$ closest to 0, $\frac{1}{2}$, or 1?

Graph $\frac{2}{12}$ on a number line. Mark off 12 equal increments from 0 to 1. You know $\frac{1}{2} = \frac{6}{12}$.

Is it time for lunch?

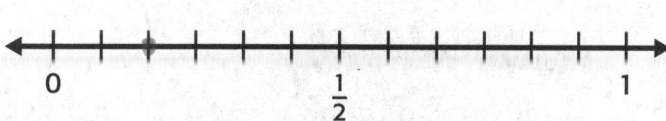

Which number, 0, $\frac{1}{2}$, or 1 is $\frac{2}{12}$ closest to on the number line? ___0___

So, the length of a poison dart frog is closest to ___0___ feet.

Key Concept Rounding Fractions

Round Down	**Round to $\frac{1}{2}$**	**Round Up**
If the numerator is much smaller than the denominator, round the fraction down to 0.	If the numerator is about half of the denominator, round the fraction to $\frac{1}{2}$.	If the numerator is almost as large as the denominator, round the fraction up to 1.

$\frac{1}{10}$ rounds to 0.	$\frac{6}{10}$ rounds to $\frac{1}{2}$.	$\frac{9}{10}$ rounds to 1.

Online Content at connectED.mcgraw-hill.com

Example 2

Round $\frac{4}{9}$ to 0, $\frac{1}{2}$, or 1.

The number $4\frac{1}{2}$ is half of 9.

The numerator in $\frac{4}{9}$ is ___4___,
which is very close to $4\frac{1}{2}$.

So, $\frac{4}{9}$ rounds to $\dfrac{1}{2}$.

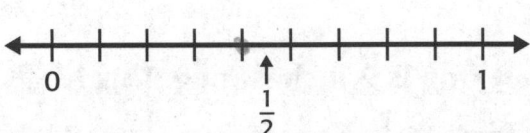

Helpful Hint

The denominator names the number of units. When rounding $\frac{4}{9}$, the number line should be divided into 9 equal sections.

Check Graph $\frac{4}{9}$ on a number line.

0     $\frac{1}{2}$ 1

$\frac{1}{2}$ is halfway between $\frac{4}{9}$ and $\frac{5}{9}$ since half of 9 is $4\frac{1}{2}$.

Guided Practice

Graph each fraction on the number line. Then state whether each fraction is closest to 0, $\frac{1}{2}$, or 1.

1. $\frac{5}{6} \approx$ 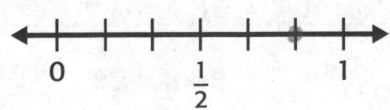 1

0     $\frac{1}{2}$ 1

Talk MATH

Tell how to round fractions in your own words.

2. $\frac{5}{8} \approx$ 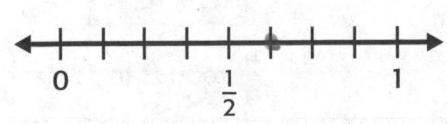 $\frac{1}{2}$

0     $\frac{1}{2}$ 1

Independent Practice

Round each fraction to 0, $\frac{1}{2}$, or 1. Use a number line if needed.

3. $\frac{1}{8} \approx$ ___0___

4. $\frac{5}{9} \approx$ ___$\frac{1}{2}$___

5. $\frac{7}{8} \approx$ ___1___

6. $\frac{3}{7} \approx$ ___$\frac{1}{2}$___

7. $\frac{5}{11} \approx$ ___$\frac{1}{2}$___

8. $\frac{4}{5} \approx$ ___1___

9. $\frac{1}{9} \approx$ ___0___

10. $\frac{6}{7} \approx$ ___1___

11. $\frac{2}{5} \approx$ ___$\frac{1}{2}$___

12. $\frac{3}{8} \approx$ ___$\frac{1}{2}$___

13. $\frac{1}{5} \approx$ ___0___

14. $\frac{15}{16} \approx$ ___1___

Problem Solving

15. **Mathematical PRACTICE** 6 **Be Precise** Round the length of the ribbon to 0 inches, $\frac{1}{2}$ inch, or 1 inch.

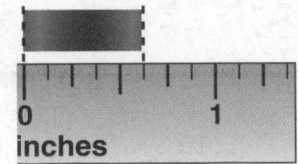

inches

$\frac{1}{2}$

16. Corri has finished about $\frac{3}{5}$ of her daily chores. Has she finished about half of her chores or almost all of them?

Half

17. Peter has read about $\frac{12}{15}$ of his book. Has he read about half of his book or almost all of his book?

almost all

HOT Problems

18. **Mathematical PRACTICE** 3 **Which One Doesn't Belong?** Circle the fraction that does not belong with the other three. Explain your reasoning.

$\frac{2}{11}$ $\frac{8}{15}$ $\frac{7}{13}$ $\frac{5}{12}$

19. 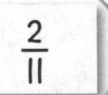 **Building on the Essential Question** Describe two different ways of rounding fractions.

You can round down to 0 or you can round up to 1.

MY Homework

Lesson 1
Round Fractions

Homework Helper

Need help? 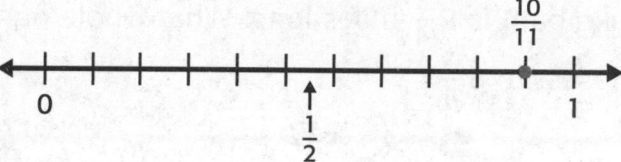 connectED.mcgraw-hill.com

Round $\frac{10}{11}$ to 0, $\frac{1}{2}$, or 1.

Compare the numerator and denominator. The numerator, 10, is close to 11.

So, $\frac{10}{11}$ rounds to 1.

Check Graph $\frac{10}{11}$ on a number line.

$\frac{1}{2}$ is halfway between $\frac{5}{11}$ and $\frac{6}{11}$ since half of 11 is $5\frac{1}{2}$.

Practice

Round each fraction to 0, $\frac{1}{2}$, or 1. Use a number line if needed.

1. $\frac{5}{9} \approx$ _$\frac{1}{2}$_

2. $\frac{1}{14} \approx$ _0_

3. $\frac{12}{13} \approx$ _1_

4. $\frac{2}{13} \approx$ _0_

5. $\frac{9}{11} \approx$ _1_

6. $\frac{9}{17} \approx$ _$\frac{1}{2}$_

 # Problem Solving

7. Kevin ate $\frac{5}{12}$ of a pizza. Which is a better estimate for the amount of pizza that he ate: about half of the pizza or almost all of the pizza?

about half of the pizza

8. **Mathematical** **PRACTICE 1** **Make Sense of Problems** Darius has mowed $\frac{1}{5}$ of his backyard. Which is a better estimate for how much of the lawn he has left to mow: almost all of the lawn or about half of the lawn?

almost all of the lawn

9. A bike path is $6\frac{3}{4}$ miles long. What whole number is closest to $6\frac{3}{4}$?

_7_

10. Savannah is making a quilt with squares having side lengths that are $\frac{15}{16}$ foot each. Are the side lengths of the squares closer to $\frac{1}{2}$ foot or 1 foot long?

1 foot long

Test Practice

11. Samantha shaded $\frac{3}{7}$ of her design.

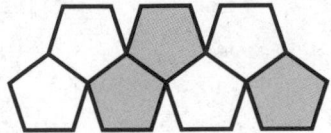

A pattern for success!

Which number is the best estimate for the shaded part of her design?

Ⓐ 0

Ⓒ $\frac{1}{2}$

Ⓑ $\frac{1}{7}$

Ⓓ 1

Add Like Fractions

Lesson 2

ESSENTIAL QUESTION
How can equivalent fractions help me add and subtract fractions?

Like fractions have the same denominator.

 **Math in My World** Watch Tools  Tutor

Example 1

The length across the bell, or top, of a mushroom jellyfish is about $\frac{5}{6}$ foot. If two mushroom jellyfish were placed side by side, what would be the combined length?

Find $\frac{5}{6} + \frac{5}{6}$.

$\frac{5}{6}$ ← → $\frac{5}{6}$

One Way Use models.

Place two sets of five $\frac{1}{6}$-tiles side by side.

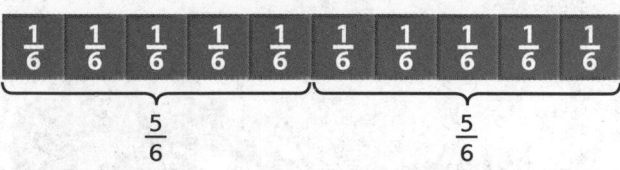

$\frac{5}{6}$ $\frac{5}{6}$

There are _____ $\frac{1}{6}$-tiles altogether.

This shows the fraction $\frac{10}{6}$, or $1\frac{2}{3}$.

Another Way Add the numerators. Keep the denominator.

$$\frac{5}{6} + \frac{5}{6} = \frac{5+5}{6}$$

$$= \frac{10}{6} \qquad 5 + 5 = 10$$

$$= 1\frac{2}{3} \qquad \text{Write as a mixed number in simplest form.}$$

So, $\frac{5}{6} + \frac{5}{6} = $ 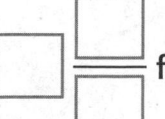.

The combined length would be ☐☐ feet.

Example 2

The table shows how much of a book Teddy read each day. What fraction of the book did Teddy read altogether on Tuesday and Thursday?

Day	Fraction Read
Monday	$\frac{1}{10}$
Tuesday	$\frac{4}{10}$
Wednesday	$\frac{3}{10}$
Thursday	$\frac{2}{10}$

On Tuesday, Teddy read $\dfrac{\Box}{\Box}$ of the book.

On Thursday, he read $\dfrac{\Box}{\Box}$ of the book.

Add the numerators. Keep the denominator.

$$\frac{4}{10} + \frac{2}{10} = \frac{4+2}{10}$$

$$= \frac{\Box}{10} \qquad \text{Add. } 4 + 2 = 6$$

$$= \frac{\Box}{5} \qquad \text{Write in simplest form.}$$

Helpful Hint

Follow these steps to simplify $\frac{6}{10}$.

$$\frac{6}{10} = \frac{6 \div 2}{10 \div 2}$$
$$= \frac{3}{5}$$

So, Teddy read $\dfrac{\Box}{\Box}$ of the book on Tuesday and Thursday.

Guided Practice

Add. Write each sum in simplest form.

Talk MATH

Describe a real-world problem that can be solved by adding like fractions.

1. $\dfrac{1}{7} + \dfrac{3}{7} = \dfrac{\Box}{\Box}$

2. $\dfrac{1}{4} + \dfrac{1}{4} = \dfrac{\Box}{\Box} = \dfrac{\Box}{\Box}$

Independent Practice

Add. Write each sum in simplest form.

3. $\frac{1}{6} + \frac{1}{6} =$ _____

4. $\frac{5}{8} + \frac{3}{8} =$ _____

5. $\frac{2}{9} + \frac{3}{9} =$ _____

6. $\frac{4}{7} + \frac{2}{7} =$ _____

7. $\frac{2}{6} + \frac{2}{6} =$ _____

8. $\frac{2}{10} + \frac{5}{10} =$ _____

9. $\frac{3}{8} + \frac{1}{8} =$ _____

10. $\frac{3}{4} + \frac{1}{4} =$ _____

11. $\frac{4}{9} + \frac{5}{9} =$ _____

Algebra Find each unknown.

12. $\frac{1}{3} + \blacksquare = \frac{2}{3}$

The unknown is _____.

13. $\frac{5}{12} + \frac{4}{12} = \frac{\blacksquare}{4}$

The unknown is _____.

14. $\frac{3}{10} + \frac{4}{\blacksquare} = \frac{7}{10}$

The unknown is _____.

Problem Solving

15. Terri painted $\frac{5}{12}$ of a fence. Rey painted $\frac{4}{12}$ of the fence. How much of the fence did they paint altogether? Write in simplest form.

_____

16. Meagan walked $\frac{4}{10}$ mile to the park. She walked the same distance home. How much did she walk altogether? Write in simplest form.

_____

17. **Mathematical PRACTICE 2** **Use Number Sense** It rained $\frac{2}{8}$ inch in one hour. It rained $\frac{4}{8}$ inch in the next hour. Find the total amount of rain. Write in simplest form.

_____

HOT Problems

18. **Mathematical PRACTICE 3** **Justify Conclusions** Select two fractions whose sum is $\frac{3}{4}$ and whose denominators are both the same, but not equal to 4. Justify your selection.

_____

19. **Building on the Essential Question** How are equivalent fractions used in writing a sum of two like fractions in simplest form?

_____

MY Homework

Homework Helper

Need help? ↗ connectED.mcgraw-hill.com

Find $\frac{3}{10} + \frac{3}{10}$. Write the sum in simplest form.

$$\frac{3}{10} + \frac{3}{10} = \frac{3+3}{10}$$ Add the numerators. Keep the denominator.

$$= \frac{6}{10}$$ Add. $3 + 3 = 6$

$$= \frac{3}{5}$$ Write in simplest form.

So, $\frac{3}{10} + \frac{3}{10} = \frac{3}{5}$.

Check The models show that $\frac{3}{10} + \frac{3}{10} = \frac{6}{10}$, or $\frac{3}{5}$.

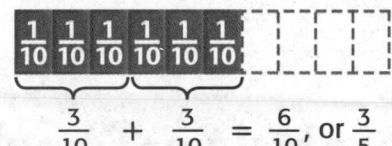

$$\frac{3}{10} + \frac{3}{10} = \frac{6}{10}, \text{ or } \frac{3}{5}$$

Practice

Add. Write each sum in simplest form.

1. $\frac{7}{10} + \frac{2}{10} =$ _____

2. $\frac{13}{16} + \frac{2}{16} =$ _____

3. $\frac{4}{5} + \frac{1}{5} =$ _____

4. $\frac{7}{15} + \frac{2}{15} =$ _____

5. $\frac{9}{20} + \frac{3}{20} =$ _____

6. $\frac{5}{8} + \frac{1}{8} =$ _____

Problem Solving

The table gives the fraction of each type of parade float used in a recent parade. Use the table to answer Exercises 7 and 8.

Type of Parade Float	Fraction
Sports Team	$\frac{6}{18}$
Radio Station	$\frac{5}{18}$
High School	$\frac{3}{18}$
Dance Group	$\frac{4}{18}$

7. What fraction of the floats were from either a dance group or a radio station? Write in simplest form.

_____

8. What fraction of the floats were not from a sports team? Write in simplest form.

_____

9. **Mathematical**  **PRACTICE** **3** **Draw a Conclusion** Sherry was in charge of distributing 25 food items that were donated to the local food pantry. On Monday, she distributed 8 items. On Tuesday, she distributed 7 items. Five more items were distributed on Wednesday. What fraction of the food items were distributed by the end of the day on Wednesday?

_____

Vocabulary Check

Complete the sentence with the correct vocabulary word(s).

10. Both fractions in the expression $\frac{1}{3} + \frac{1}{3}$ are examples of _____.

Test Practice

11. Gina is working on a jigsaw puzzle. She completed $\frac{1}{10}$ of the puzzle yesterday and $\frac{3}{10}$ of the puzzle today. In simplest form, what fraction of the puzzle is completed?

(A) $\frac{2}{5}$ (C) $\frac{2}{10}$

(B) $\frac{3}{5}$ (D) $\frac{3}{10}$

We're a great fit!

Subtract Like Fractions

Lesson 3

ESSENTIAL QUESTION
How can equivalent fractions help me add and subtract fractions?

 **Math in My World**   Watch   Tools   Tutor

We're out of this world!

Example 1

About $\frac{7}{10}$ of Earth's surface is covered by oceans. The Pacific Ocean is the largest ocean and covers about $\frac{3}{10}$ of Earth's surface. How much of Earth's surface is covered by oceans other than the Pacific Ocean?

Find $\frac{7}{10} - \frac{3}{10}$.

One Way Use Models.

Place seven $\frac{1}{10}$-fraction tiles.

$$\frac{1}{10} \ \frac{1}{10} \ \frac{1}{10} \ \frac{1}{10} \ \cancel{\ } \ \cancel{\ } \ \cancel{\ }$$

Remove three of the tiles.

There are _____ tiles left, which represent $\frac{4}{10}$, or $\frac{2}{5}$.

So, $\dfrac{\boxed{}}{\boxed{}}$ of Earth's surface is covered by oceans other than the Pacific Ocean.

Another Way Subtract the numerators. Keep the denominator.

$$\frac{7}{10} - \frac{3}{10} = \frac{7 - 3}{10}$$

$$= \frac{4}{10} \qquad 7 - 3 = 4$$

$$= \dfrac{\boxed{}}{\boxed{}} \qquad \text{Write in simplest form.}$$

Example 2

The table shows the amount of rainfall several cities received in a recent month. How much more rain did Centerville receive than Brushton? Write in simplest from.

City	Rainfall (in.)
Spring Valley	$\frac{1}{10}$
Clarksburg	$\frac{6}{10}$
Centerville	$\frac{9}{10}$
Brushton	$\frac{3}{10}$

Subtract the numerators. Keep the denominator the same.

$$\frac{9}{10} - \frac{3}{10} = \frac{9-3}{10}$$

$$= \frac{6}{10} \qquad 9 - 3 = 6$$

$$= \frac{\square}{\square} \qquad \text{Write in simplest form.}$$

So, $\frac{\square}{\square}$ inch more rain fell in _____ than in _____ .

Guided Practice

Subtract. Write each difference in simplest form.

1. $\frac{5}{7} - \frac{3}{7} = \frac{\square}{\square}$

2. $\frac{3}{5} - \frac{2}{5} = \frac{\square}{\square}$

3. $\frac{6}{9} - \frac{3}{9} = \frac{\square}{\square}$

Talk MATH

Tell about a real-world situation in which you would find $\frac{3}{4} - \frac{1}{4}$.

Name _____

Independent Practice

Subtract. Write each difference in simplest form.

4. $\dfrac{5}{6} - \dfrac{3}{6} =$ ___ $\dfrac{1}{3}$

5. $\dfrac{2}{3} - \dfrac{1}{3} =$ ___ $\dfrac{1}{3}$

6. $\dfrac{3}{5} - \dfrac{1}{5} =$ ___ $\dfrac{2}{5}$

7. $\dfrac{6}{7} - \dfrac{5}{7} =$ ___ $\dfrac{1}{7}$

8. $\dfrac{5}{9} - \dfrac{2}{9} =$ ___ $\dfrac{3}{9}$

9. $\dfrac{6}{8} - \dfrac{4}{8} =$ ___ $\dfrac{2}{8}$

10. $\dfrac{3}{4} - \dfrac{1}{4} =$ ___ $\dfrac{2}{4}$

11. $\dfrac{9}{12} - \dfrac{3}{12} =$ ___ $\dfrac{6}{12}$

12. $\dfrac{4}{5} - \dfrac{2}{5} =$ ___ $\dfrac{2}{5}$

Algebra Find each unknown.

13. $\dfrac{5}{9} - \dfrac{1}{9} = b$

$b =$ ___ $\dfrac{4}{9}$

14. $\dfrac{6}{8} - \dfrac{h}{8} = \dfrac{1}{8}$

$h =$ ___ 5

15. $\dfrac{3}{4} - \dfrac{2}{4} = \dfrac{w}{4}$

$w =$ ___ 1

 # Problem Solving

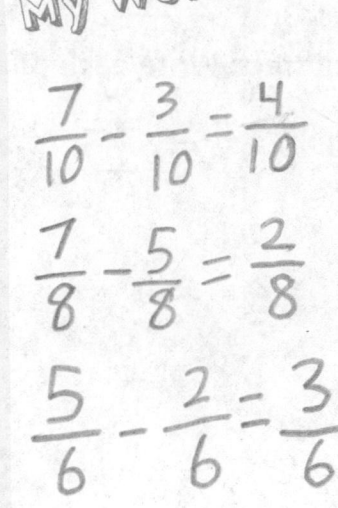

16. **Mathematical**
PRACTICE 1 **Make Sense of Problems** A bucket was $\frac{7}{10}$ full with water. After Vick washed the car, the bucket was only $\frac{3}{10}$ full. What fraction of the water in the bucket did Vick use to wash the car? Write in simplest form.

$$\frac{4}{10} \quad \frac{2}{5}$$

17. Roshanda bought $\frac{5}{8}$ pound of ham and $\frac{7}{8}$ pound of roast beef. How much more roast beef than ham did she buy? Write in simplest form.

$$\frac{2}{8} \quad \frac{1}{4}$$

18. Chris spent $\frac{5}{6}$ hour drawing and $\frac{2}{6}$ hour reading. How much more time did he spend drawing than reading? Write in simplest form.

$$\frac{3}{6} \quad \frac{1}{2}$$

My Work! handwritten:
$$\frac{7}{10} - \frac{3}{10} = \frac{4}{10}$$
$$\frac{7}{8} - \frac{5}{8} = \frac{2}{8}$$
$$\frac{5}{6} - \frac{2}{6} = \frac{3}{6}$$

HOT Problems

19. **Mathematical**
PRACTICE 2 **Use Number Sense** Felisa wrote the equation $\frac{5}{6} - \frac{2}{3} = \frac{2}{3}$. How can you tell her answer is incorrect without calculating?

Her answer is wrong in two ways, one the denominator is not the same two is 5-2 does not equal 2.

20. **Building on the Essential Question** How can models and equations help me subtract like fractions?

MY Homework

Homework Helper

Need help? connectED.mcgraw-hill.com

Carla volunteers at a veterinarian's office. Of her time spent there, $\frac{6}{10}$ is spent grooming dogs and $\frac{1}{10}$ is spent answering phone calls. How much more of her time is spent grooming dogs than answering phone calls? Write the difference in simplest form.

Find $\frac{6}{10} - \frac{1}{10}$.

$\frac{6}{10} - \frac{1}{10} = \frac{6-1}{10}$ Subtract the numerators. Keep the denominator the same.

$= \frac{5}{10}$ $6 - 1 = 5$

$= \frac{1}{2}$ Write in simplest form.

So, $\frac{6}{10} - \frac{1}{10} = \frac{1}{2}$.

Carla spends $\frac{1}{2}$ more of her time grooming dogs than answering phone calls.

Check The models show that $\frac{6}{10} - \frac{1}{10} = \frac{5}{10}$ or $\frac{1}{2}$.

| $\frac{1}{10}$ | $\frac{1}{10}$ | $\frac{1}{10}$ | $\frac{1}{10}$ | $\frac{1}{10}$ | ✗ | | | | |

Practice

Subtract. Write each difference in simplest form.

1. $\frac{3}{6} - \frac{1}{6} =$ _____

2. $\frac{7}{9} - \frac{3}{9} =$ _____

3. $\frac{7}{8} - \frac{2}{8} =$ _____

Problem Solving

The table shows the results of a survey of 28 students and their favorite tourist attractions. Use the table to answer Exercises 4 and 5.

Place	Fraction of Students
Mt. Rushmore	$\frac{14}{28}$
Grand Canyon	$\frac{8}{28}$
Statue of Liberty	$\frac{6}{28}$

4. What fraction of students prefer Mt. Rushmore over the Grand Canyon? Write in simplest form.

_____

5. Suppose four students change their minds and choose the Statue of Liberty instead of the Grand Canyon. What part of the class now prefers Mt. Rushmore over the State of Liberty? Write in simplest form.

_____

Mathematical
6. PRACTICE **1** **Make Sense of Problems** On a class trip to the museum, $\frac{5}{8}$ of the students saw the dinosaurs and $\frac{2}{8}$ of the students saw the jewelry collection. What fraction more of students saw the dinosaurs than the jewelry collection? Write in simplest form.

_____

7. The Indian Ocean is $\frac{2}{10}$ of the area of the world's oceans. What fraction represents the area of the remaining oceans that make up the world's oceans? Write in simplest form.

_____

Test Practice

8. The pictures to the right show how much sausage and pepperoni pizza was left at the end of one day. Which fraction represents how much more sausage pizza than pepperoni pizza was left?

Sausage   Pepperoni

(A) $\frac{7}{8}$ (C) $\frac{3}{8}$

(B) $\frac{4}{8}$ (D) $\frac{11}{8}$

Hands On
Use Models to Add Unlike Fractions

Lesson 4

ESSENTIAL QUESTION
How can equivalent fractions help me add and subtract fractions?

Unlike fractions have different denominators. Before you can add unlike fractions, one or both of the fractions must be renamed so that they have a common denominator.

Build It  Tools

To finish building a birdhouse, Jordan uses two boards. One is $\frac{1}{2}$ foot long and the other is $\frac{1}{4}$ foot long. What is the total length of the boards?

Bird at work!

1 Model each fraction using fraction tiles and place them side by side.

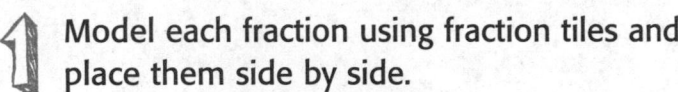

| $\frac{1}{2}$ | $\frac{1}{4}$ |

| $\frac{1}{4}$ | $\frac{1}{4}$ | $\frac{1}{4}$ |

2 Find fraction tiles that will match the length of the combined tiles. Line them up below the model.

3 Count. There are _____ of the $\frac{1}{4}$-fraction tiles in all. This represents the fraction $\frac{3}{4}$.

So, $\frac{1}{2} + \frac{1}{4} = \dfrac{\boxed{}}{\boxed{}}$. The total length of the boards is $\dfrac{\boxed{}}{\boxed{}}$ foot.

Try It

Muna's family ate $\frac{2}{3}$ of a strawberry pie and Brendan's family ate $\frac{3}{4}$ of a different strawberry pie. How much did they eat altogether?

1 Model each fraction using fraction tiles and place them side by side.

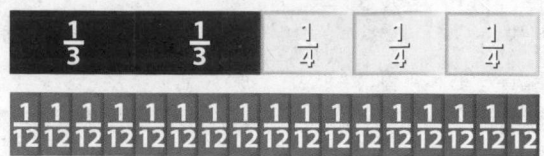

2 Find fraction tiles that will match the length of the combined tiles. Line them up below the model.

3 Count. There are _____ of the $\frac{1}{12}$-fraction tiles in all. This represents the fraction $\frac{17}{12}$, or $1\frac{5}{12}$.

So, $\frac{2}{3} + \frac{3}{4} = 1\frac{5}{12}$.

They ate $1\frac{5}{12}$ strawberry pies altogether.

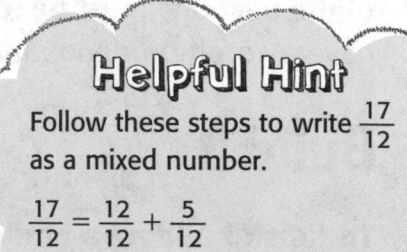

Helpful Hint
Follow these steps to write $\frac{17}{12}$ as a mixed number.

$$\frac{17}{12} = \frac{12}{12} + \frac{5}{12}$$
$$= 1\frac{5}{12}$$

Talk About It

1. In the first activity, how does the denominator of the sum, $\frac{3}{4}$, compare to the denominators of the addends, $\frac{1}{2}$ and $\frac{1}{4}$?

_____

_____

2. In the second activity, how does the denominator of the sum, $1\frac{5}{12}$, compare to the denominators of the addends, $\frac{2}{3}$ and $\frac{3}{4}$?

_____

_____

3. **Mathematical PRACTICE 6** **Explain to a Friend** Use your answers from Exercises 1 and 2 to predict the denominator of the sum of $\frac{1}{3}$ and $\frac{1}{4}$. Explain.

_____

_____

Practice It

Find the sum using fraction tiles. Write in simplest form.
Draw the models.

4. $\dfrac{2}{3} + \dfrac{1}{6} =$ _____

5. $\dfrac{3}{8} + \dfrac{1}{4} =$ _____

6. $\dfrac{3}{10} + \dfrac{1}{5} =$ _____

7. $\dfrac{5}{8} + \dfrac{1}{4} =$ _____

8. $\dfrac{1}{3} + \dfrac{1}{4} =$ _____

9. $\dfrac{3}{4} + \dfrac{1}{6} =$ _____

 **Apply It**

Mathematical PRACTICE 5 **Use Math Tools** Draw models to solve
Exercises 10–12.

 My Drawing!

10. Alana walks east $\frac{4}{9}$ mile to school every morning.
On Saturday, she walked to a friend's house, which is
$\frac{1}{3}$ mile farther east from her school. How far did
Alana walk to her friend's house on Saturday?

_____

11. Mr. Hawkins gets two bonus payments every year. This
year, his first bonus was $\frac{1}{4}$ of his salary. His second bonus
was $\frac{1}{8}$ of his salary. What was the fraction of his salary for
his combined bonus payments?

12. At a gas station, Kurt asked for directions to the nearest
town. The attendant told him to go $\frac{5}{6}$ mile south
and then $\frac{1}{4}$ mile east. How far does Kurt have to
drive to get to the town?

_____

 Which way do I go?

Mathematical PRACTICE 4
13. **Model Math** Write a real-world problem
that can be solved by adding unlike fractions.

_____

_____

Write About It

14. How can I use models to add fractions?

_____

_____

_____

MY Homework

Homework Helper eHelp

Need help? ↗ connectED.mcgraw-hill.com

Find the sum of $\frac{2}{3}$ and $\frac{1}{2}$.

1 Model each fraction using fraction tiles and place them side by side.

$\frac{2}{3}$

$\frac{1}{3}$	$\frac{1}{3}$	$\frac{1}{2}$				
$\frac{1}{6}$	$\frac{1}{6}$	$\frac{1}{6}$	$\frac{1}{6}$	$\frac{1}{6}$	$\frac{1}{6}$	$\frac{1}{6}$

$\frac{7}{6}$

2 Find fraction tiles that will match the length of the combined tiles. Line them up below the model.

3 Count. There are seven of the $\frac{1}{6}$-fraction tiles in all.

Helpful Hint

Follow these steps to write $\frac{7}{6}$ as a mixed number.

$$\frac{7}{6} = \frac{6}{6} + \frac{1}{6} = 1\frac{1}{6}$$

So, $\frac{2}{3} + \frac{1}{2} = \frac{7}{6}$ or $1\frac{1}{6}$.

Practice

Find the sum using the fraction tiles shown. Write in simplest form.

1. $\frac{1}{2} + \frac{1}{4} =$ _____

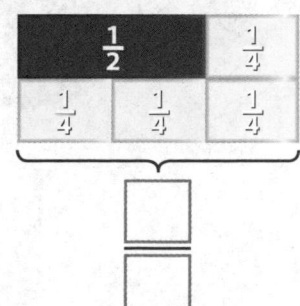

2. $\frac{1}{2} + \frac{1}{6} -$ _____

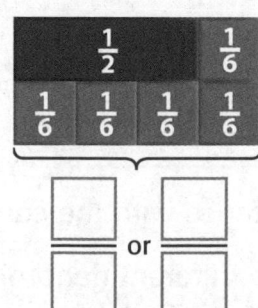

Problem Solving

Mathematical PRACTICE 5 Use Math Tools Draw models to solve Exercises 3–6. Write in simplest form.

My Drawing!

3. After school, Maurice walks $\frac{1}{3}$ mile to the park and then walks $\frac{1}{2}$ mile to his house. How far does Maurice walk from school to his house?

4. Ricki took a survey in the fifth grade and found that $\frac{2}{3}$ of the students ride the bus to school, and $\frac{1}{4}$ of the students walk. What fraction of the fifth grade students either ride the bus or walk to school?

5. Elizabeth made an English muffin pizza using $\frac{1}{4}$ cup of cheese and $\frac{3}{8}$ cup of sausage. How many cups of toppings did she use?

6. Craig and Alissa are building a sandcastle on the beach. They each have a bucket. Craig's bucket holds $\frac{1}{2}$ pound of sand and Alissa's bucket holds $\frac{9}{10}$ pound of sand. How much sand can Craig and Alissa collect together at one time?

Vocabulary Check

7. Complete the sentence with the correct vocabulary word(s).

Fractions that have different denominators are called _____.

Add Unlike Fractions

 **Math in My World**

Example 1

In the morning, an octopus swam for $\frac{1}{3}$ hour.

In the afternoon, the octopus swam for $\frac{1}{4}$ hour. For how much of one hour did the octopus swim altogether?

Find $\frac{1}{3} + \frac{1}{4}$.

Write equivalent, like fractions using the least common denominator, LCD. The LCD of $\frac{1}{3}$ and $\frac{1}{4}$ is 12.

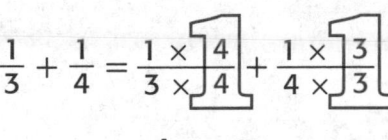 Write equivalent fractions using the LCD.

$= \frac{4}{12} + \frac{3}{12}$ Multiply.

$= \frac{4+3}{12}$, or □   Add like fractions.

> **Helpful Hint**
> The least common denominator, LCD, is the least common multiple of the denominators.

So, $\frac{1}{3} + \frac{1}{4} = \dfrac{□}{□}$. The octopus swam for $\dfrac{□}{□}$ hour altogether.

Check The models show that $\frac{1}{3} + \frac{1}{4} = \dfrac{□}{□}$.

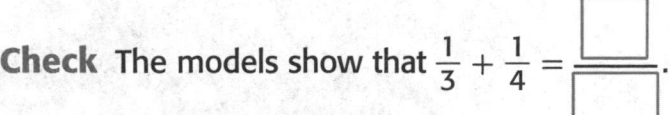

You can use benchmark fractions such as $\frac{1}{2}$ to check an answer for reasonableness.

Example 2

Catalina spent $\frac{1}{5}$ of her free time reading and $\frac{7}{10}$ of her free time practicing her flute. What fraction of her free time did she spend reading and practicing her flute?

Add $\frac{1}{5}$ and $\frac{7}{10}$.

Estimate $\frac{1}{5} + \frac{7}{10} \approx 0 + 1$, or 1

Write equivalent, like fractions using the LCD. The LCD is 10.

$$\frac{1}{5} + \frac{7}{10} = \frac{1 \times 2}{5 \times 2} + \frac{7}{10}$$

$$= \frac{\square}{10} + \frac{7}{10} \qquad \text{Multiply.}$$

$$= \frac{\square + 7}{10}, \text{ or } \frac{\square}{10} \qquad \text{Add like fractions.}$$

Catalina spent $\frac{\square}{\square}$ of her free time reading and practicing her flute.

Check $\frac{9}{10} \approx 1$

How can benchmark fractions and number sense be used to check answers for reasonableness?

Guided Practice

Add. Write each sum in simplest form.

1. $\frac{2}{5} + \frac{1}{2} = \frac{\square}{\square}$

2. $\frac{3}{4} + \frac{1}{8} = \frac{\square}{\square}$

Independent Practice

Add. Write each sum in simplest form.

3. $\dfrac{1}{3} + \dfrac{1}{5} =$ _____

4. $\dfrac{1}{2} + \dfrac{1}{5} =$ _____

5. $\dfrac{5}{12} + \dfrac{1}{4} =$ _____

6. $\dfrac{2}{3} + \dfrac{1}{6} =$ _____

7. $\dfrac{1}{2} + \dfrac{1}{4} =$ _____

8. $\dfrac{5}{8} + \dfrac{1}{16} =$ _____

9. $\dfrac{3}{5} + \dfrac{3}{10} =$ _____

10. $\dfrac{5}{8} + \dfrac{3}{16} =$ _____

11. $\dfrac{3}{5} + \dfrac{3}{20} =$ _____

Algebra Find each unknown.

12. $\dfrac{7}{12} + \dfrac{1}{3} = x$

13. $\dfrac{3}{10} + \dfrac{3}{0} = \dfrac{9}{y}$

14. $\dfrac{3}{0} + \dfrac{2}{5} = \dfrac{w}{40}$

$x =$ _____

$y =$ _____

$w =$ _____

Problem Solving

15. A farmer harvested some of his pecan crop on Friday and Saturday. What fraction of the pecan crop was harvested in the two days?

Pecan Harvest	
Day	**Amount**
Friday	$\frac{3}{8}$
Saturday	$\frac{1}{3}$

My Work!

_____

16. Angel has two chores after school. She rakes leaves for $\frac{3}{4}$ hour and spends $\frac{1}{2}$ hour washing the car. How long does Angel spend on her chores in all?

_____

Mathematical
17. PRACTICE 2 Use Number Sense Leon found the sum of $\frac{5}{6}$ and $\frac{2}{3}$ to be $\frac{11}{12}$. How can you tell that his answer is incorrect without calculating?

_____

_____

Mathematical
18. PRACTICE 3 Which One Doesn't Belong? Circle the expression that does not belong with the other three. Explain your reasoning.

$$\frac{5}{6} + \frac{1}{3} \qquad \frac{5}{6} + \frac{1}{2} \qquad \frac{5}{6} + \frac{2}{6} \qquad \frac{5}{6} + \left(\frac{1}{6} + \frac{1}{6}\right)$$

_____

19. **? Building on the Essential Question** How are equivalent fractions used when adding unlike fractions?

_____

_____

MY Homework

Lesson 5

Add Unlike Fractions

Homework Helper  eHelp

Need help? connectED.mcgraw-hill.com

Find $\frac{1}{6} + \frac{1}{4}$.

Write equivalent, like fractions using the least common

denominator, LCD. The LCD of $\frac{1}{6}$ and $\frac{1}{4}$ is 12.

$\frac{1}{6} + \frac{1}{4} = \frac{1 \times 2}{6 \times 2} + \frac{1 \times 3}{4 \times 3}$ Write equivalent fractions using the LCD.

$= \frac{2}{12} + \frac{3}{12}$ Multiply.

$= \frac{2+3}{12}$, or $\frac{5}{12}$ Add like fractions.

So, $\frac{1}{6} + \frac{1}{4} = \frac{5}{12}$.

Check The models show that $\frac{1}{6} + \frac{1}{4} = \frac{5}{12}$.

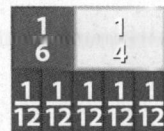

Practice

Add. Write each sum in simplest form.

1. $\frac{5}{8} + \frac{3}{10} =$ _____

2. $\frac{3}{5} + \frac{1}{4} =$ _____

3. $\frac{4}{7} + \frac{1}{8} =$ _____

Problem Solving

4. Tashia ate $\frac{1}{3}$ of a pizza, and Jay ate $\frac{3}{8}$ of the same pizza. What fraction of the pizza was eaten?

_____

5. Basir took a science test on Friday. One-eighth of the questions were multiple choice, and $\frac{3}{4}$ of the questions were true-false questions. What part of the total number of questions are either multiple choice or true-false questions?

_____

Mathematical
6. **PRACTICE 2** **Use Number Sense** Edison delivers $\frac{1}{5}$ of the newspapers in the neighborhood, and Anita delivers $\frac{1}{2}$ of them. Together, Edison and Anita deliver what fraction of the newpapers?

_____

7. Dylan and Sonia are hiking different trails. If Dylan hiked Riverwalk and Mountainview, and Sonia hiked Mountainview and Pine, how many miles did each of them hike?

Hiking Trails	
Trail	**Distance (mi)**
Riverwalk	$\frac{3}{4}$
Mountainview	$\frac{1}{2}$
Pine	$\frac{3}{5}$

Test Practice

8. Which expression will have the same sum as $\frac{3}{8} + \frac{1}{4}$?

Ⓐ $\frac{3}{8} + \frac{1}{8}$

Ⓒ $\frac{3}{4} + \frac{1}{4}$

Ⓑ $\left(\frac{1}{8} + \frac{1}{8} + \frac{1}{8}\right) + \frac{1}{4}$

Ⓓ $\left(\frac{1}{8} + \frac{1}{8}\right) + \frac{1}{8}$

Vocabulary Check

Write the vocabulary words that describes each set of fractions below.

like fractions   unlike fractions

1. $\frac{1}{2}$ and $\frac{1}{3}$ _____   **2.** $\frac{2}{9}$ and $\frac{2}{9}$ _____

Concept Check

Round each fraction to 0, $\frac{1}{2}$, or 1. Use a number line if needed.

3. $\frac{9}{16} \approx$ _____   **4.** $\frac{12}{15} \approx$ _____   **5.** $\frac{1}{10} \approx$ _____

Add. Write each sum in simplest form.

6. $\frac{5}{9} + \frac{1}{9} =$ _____   **7.** $\frac{1}{7} + \frac{6}{7} =$ _____   **8.** $\frac{7}{16} + \frac{3}{16} =$ _____

9. $\frac{1}{6} + \frac{7}{12} =$ _____   **10.** $\frac{1}{8} + \frac{3}{4} =$ _____   **11.** $\frac{1}{4} + \frac{3}{8} =$ _____

Subtract. Write each difference in simplest form.

12. $\frac{4}{5} - \frac{1}{5} =$ _____   **13.** $\frac{11}{20} - \frac{7}{20} =$ _____   **14.** $\frac{9}{16} - \frac{7}{16} =$ _____

15. Mike spent $\frac{11}{12}$ hour talking on his cell phone. Brent spent $\frac{7}{12}$ hour talking on his cell phone. How much more time did Mike spend on his cell phone than Brent?

_____

16. Christie decided to make bracelets for the fifth grade class girls. Two-fifths of the bracelets were finished on Monday, and $\frac{3}{7}$ of the bracelets were finished on Tuesday. What part of the total number of bracelets are finished?

_____

17. Jamila ate $\frac{3}{8}$ of a salad. Manny ate $\frac{2}{8}$ of the same salad. How much salad did they eat altogether?

_____

Test Practice

18. The table shows the fraction of a book Karen read Saturday and Sunday. What fraction of her book did she read on these two days?

Ⓐ $\frac{3}{8}$ Ⓒ $\frac{8}{15}$

Ⓑ $\frac{1}{2}$ Ⓓ $\frac{11}{15}$

Day	Fraction of Book Read
Saturday	$\frac{1}{3}$
Sunday	$\frac{2}{5}$

Hands On
Use Models to Subtract Unlike Fractions

You can use fraction tiles to subtract fractions with unlike denominators.

Build It

Akio lives $\frac{4}{5}$ mile from school. Bianca lives $\frac{3}{10}$ mile from school. How much farther from school does Akio live than Bianca?

1 Model each fraction using fraction tiles.

Place the $\frac{1}{10}$-tiles below the $\frac{1}{5}$-tiles.

Akio

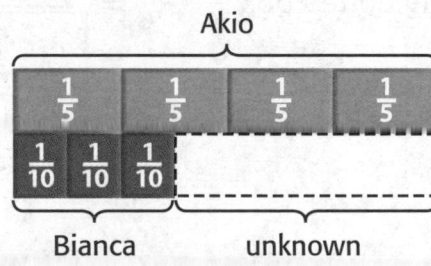

Bianca   unknown

2 Find which fraction tiles will fill in the area of the dotted box.

Try $\frac{1}{3}$-tiles. Do they fill the dotted box? _____

Try $\frac{1}{2}$-tiles. Do they fill the dotted box? _____

How many $\frac{1}{2}$-tiles fill the dotted box? _____

Since $\dfrac{\Box}{\Box}$ fills in the area of the dotted box, $\dfrac{4}{5} - \dfrac{3}{10} = \dfrac{\Box}{\Box}$.

Akio lives $\dfrac{\Box}{\Box}$ mile farther from school than Bianca.

Try It

Find $\frac{3}{4} - \frac{1}{6}$.

1 Model each fraction using fraction tiles. Place the $\frac{1}{6}$-tiles below the $\frac{1}{4}$-tiles.

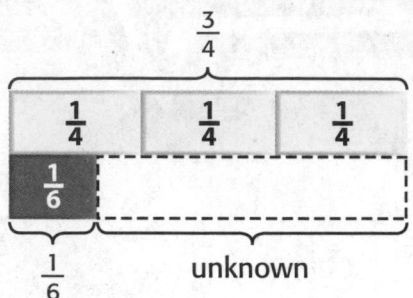

2 Find which fraction tiles will fill in the area of the dotted box.

Try $\frac{1}{3}$-tiles. Do they fill the dotted box? _____

Try $\frac{1}{12}$-tiles. Do they fill the dotted box? _____

How many $\frac{1}{12}$-tiles fill the dotted box? _____

Since $\dfrac{\square}{\square}$ fills in the area of the dotted box, $\dfrac{3}{4} - \dfrac{1}{6} = \dfrac{\square}{\square}$.

So, $\dfrac{3}{4} - \dfrac{1}{6} = \dfrac{\square}{\square}$.

Talk About It

1. Would any of the other fraction tiles fit inside the dotted box for the first activity? Explain.

_____

_____

2. 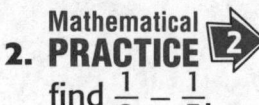 **Mathematical PRACTICE 2 Stop and Reflect** Describe how you would use fraction tiles to find $\frac{1}{2} - \frac{1}{3}$.

_____

_____

Practice It

Find each difference using fraction tiles. Draw the models.

3. $\dfrac{2}{3} - \dfrac{1}{6} =$ _____

4. $\dfrac{5}{8} - \dfrac{1}{4} =$ _____

5. $\dfrac{1}{2} - \dfrac{1}{6} =$ _____

6. $\dfrac{3}{5} - \dfrac{1}{2} =$ _____

7. $\dfrac{3}{4} - \dfrac{3}{8} =$ _____

8. $\dfrac{5}{6} - \dfrac{1}{4} =$ _____

Apply It

Mathematical
PRACTICE 5 **Use Math Tools** Draw fraction tiles to help you solve Exercises 9 and 10.

9. Missy ran $\frac{5}{8}$ mile to warm up for softball practice, while Farrah only ran $\frac{1}{2}$ mile. How much farther did Missy run?

_____

10. Pablo used $\frac{1}{2}$ of an oxygen tank while scuba diving. He used another $\frac{1}{6}$ of the tank exploring underwater. How much oxygen is left in the tank?

_____

My Drawing!

Mathematical
11. **PRACTICE 4** **Model Math** Write a real-world problem that could be represented by the fraction tiles shown.

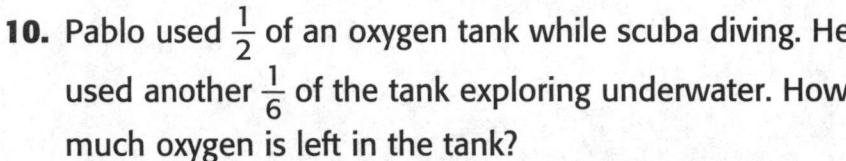

_____

_____

_____

Write About It

12. How do fraction tiles help me subtract unlike fractions?

_____

_____

MY Homework

Homework Helper eHelp

Need help? connectED.mcgraw-hill.com

Find $\frac{7}{8} - \frac{3}{4}$.

1 Model each fraction using fraction tiles.

Place the $\frac{1}{4}$-tiles below the $\frac{1}{8}$-tiles.

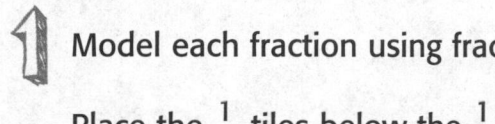

2 Find which fraction will fill in the area of the dotted box.

Try $\frac{1}{3}$-tiles. They do not fill the dotted box.

Try a $\frac{1}{8}$-tile. It does fill the dotted box.

Since $\frac{1}{8}$ fills in the area of the dotted box, $\frac{7}{8} - \frac{3}{4} = \frac{1}{8}$.

Practice

Find each difference using the fraction tiles.

1. $\frac{7}{8} - \frac{1}{2} = $ _____

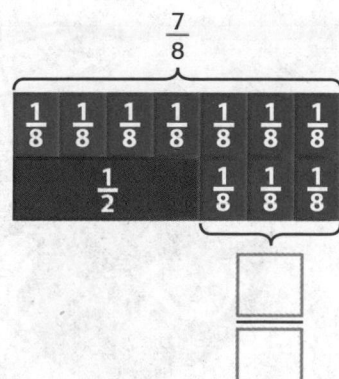

2. $\frac{2}{3} - \frac{1}{4} = $ _____

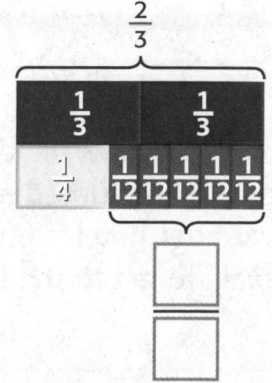

Problem Solving

3. Noah bought $\frac{1}{2}$ pound of candy to share with his friends. They ate $\frac{3}{8}$ pound of the candy. How much candy does Noah have left?

_____

4. Mr. Corwin gave his students $\frac{3}{4}$ hour to study for a test. After $\frac{1}{3}$ hour, he played a review game for the remaining time. How much time did Mr. Corwin spend playing the review game?

_____

5. Mrs. Washer filled the gas tank of her car. She used $\frac{2}{3}$ of a tank of gasoline while driving to the beach. She used another $\frac{1}{6}$ of the tank driving to her hotel. How much gasoline is left in the tank?

_____

6. Starting from her hotel, Angie walked $\frac{2}{3}$ mile along the beach in one direction. She turned around and walked $\frac{1}{2}$ mile toward her hotel. How much farther does she need to walk to get to the hotel?

_____

My Drawing!

Subtract Unlike Fractions

Lesson 7

ESSENTIAL QUESTION
How can equivalent fractions help me add and subtract fractions?

Subtracting unlike fractions is similar to adding unlike fractions.

 **Math in My World**

Island hopper!

Example 1

A female Cuban tree frog can be up to $\frac{5}{12}$ foot long. A male Cuban tree frog can be up to $\frac{1}{4}$ foot long. How much longer is the female Cuban tree frog than the male?

Find $\frac{5}{12} - \frac{1}{4}$.

Write equivalent, like fractions using the least common denominator, LCD. The LCD of $\frac{5}{12}$ and $\frac{1}{4}$ is 12.

$\frac{5}{12} - \frac{1}{4} = \frac{5}{12} - \frac{1 \times 3}{4 \times 3}$ Write equivalent fractions using the LCD.

$= \frac{5}{12} - \frac{3}{12}$ Multiply.

$= \frac{5-3}{12}$, or $\frac{\boxed{}}{\boxed{}}$ Subtract like fractions.

$= \frac{\boxed{}}{\boxed{}}$ Simplify.

A female Cuban tree frog is $\frac{\boxed{}}{\boxed{}}$ foot longer than the male.

Check for Reasonableness Use benchmark fractions to check.

Since, $\frac{1}{6} < \frac{1}{2}$, your answer is reasonable.

Example 2

Jessie finished $\frac{1}{2}$ of her homework. Lakshani finished $\frac{4}{5}$ of her homework. What fraction more of her homework did Lakshani finish than Jessie?

Portion of Homework Completed	
Jessie	$\frac{1}{2}$
Lakshani	$\frac{4}{5}$

Estimate Use benchmark fractions.

$$\frac{4}{5} - \frac{1}{2} \approx 1 - \frac{1}{2} = \frac{\boxed{}}{\boxed{}}$$

Subtract $\frac{4}{5} - \frac{1}{2}$.

Write equivalent, like fractions using the least common denominator, LCD. The LCD of $\frac{4}{5}$ and $\frac{1}{2}$ is 10.

$$\frac{4}{5} - \frac{1}{2} = \frac{4 \times \boxed{2}}{5 \times \boxed{2}} - \frac{1 \times \boxed{5}}{2 \times \boxed{5}} \qquad \text{Write equivalent fractions using the LCD.}$$

$$= \frac{8}{10} - \frac{5}{10} \qquad \text{Multiply.}$$

$$= \frac{8 - 5}{10}, \text{ or } \frac{\boxed{}}{\boxed{}} \qquad \text{Subtract like fractions.}$$

Lakshani finished $\dfrac{\boxed{}}{\boxed{}}$ more of her homework than Jessie.

Check for Reasonableness Compare to your estimate. $\dfrac{\boxed{}}{\boxed{}} \approx \frac{1}{2}$

Talk MATH

Describe the steps you can use to find $\frac{3}{4} - \frac{1}{12}$.

Guided Practice

1. Subtract. Write in simplest form.

$$\frac{3}{8} - \frac{1}{4} = \frac{\boxed{}}{\boxed{}}$$

Independent Practice

Subtract. Write each in simplest form.

2. $\dfrac{5}{6} - \dfrac{1}{2} =$ _____

3. $\dfrac{2}{5} - \dfrac{1}{4} =$ _____

4. $\dfrac{4}{5} - \dfrac{1}{6} =$ _____

5. $\dfrac{7}{8} - \dfrac{1}{2} =$ _____

6. $\dfrac{7}{12} - \dfrac{1}{3} =$ _____

7. $\dfrac{5}{6} - \dfrac{1}{3} =$ _____

8. $\dfrac{2}{3} - \dfrac{3}{10} =$ _____

9. $\dfrac{5}{8} - \dfrac{1}{2} =$ _____

10. $\dfrac{4}{5} - \dfrac{2}{15} =$ _____

Algebra Find the unknown.

11. $\dfrac{5}{6} - \dfrac{3}{4} = m$

12. $\dfrac{2}{3} - \dfrac{3}{5} = \dfrac{n}{15}$

13. $\dfrac{5}{12} - \dfrac{1}{6} = p$

$m =$ _____

$n =$ _____

$p =$ _____

 **Problem Solving**

14. Angie rides her bicycle $\frac{2}{3}$ mile to school. On Friday, she took a shortcut so that the ride to school was $\frac{1}{9}$ mile shorter. How long was Angie's bicycle ride on Friday?

15. Mathematical
PRACTICE 6 **Be Precise** Ollie used $\frac{1}{2}$ cup of vegetable oil to make brownies. She used another $\frac{1}{3}$ cup of oil to make muffins. How much more oil did she use to make brownies?

16. Danielle poured $\frac{3}{4}$ gallon of water from a $\frac{7}{8}$ gallon bucket. How much water is left in the bucket?

HOT Problems

17. Mathematical
PRACTICE 2 **Use Number Sense** Is finding $\frac{9}{10} - \frac{1}{2}$ the same as finding $\frac{9}{10} - \frac{1}{4} - \frac{1}{4}$? Explain.

scrub-a-dub-dub!

18. **Building on the Essential Question** How are equivalent fractions used when subtracting unlike fractions?

MY Homework

Homework Helper eHelp

Need help? connectED.mcgraw-hill.com

Find $\frac{2}{5} - \frac{1}{10}$.

Estimate Use benchmark fractions.

$$\frac{2}{5} - \frac{1}{10} \approx \frac{1}{2} - 0 = \frac{1}{2}$$

Subtract $\frac{2}{5} - \frac{1}{10}$.

Write equivalent, like fractions using the least common denominator, LCD. The LCD of $\frac{2}{5}$ and $\frac{1}{10}$ is 10.

$$\frac{2}{5} - \frac{1}{10} = \frac{2 \times 2}{5 \times 2} - \frac{1}{10}$$ Write equivalent fractions using the LCD.

$$= \frac{4}{10} - \frac{1}{10}$$ Multiply.

$$= \frac{4-1}{10}, \text{ or } \frac{3}{10}$$ Subtract like fractions.

So, $\frac{2}{5} - \frac{1}{10} = \frac{3}{10}$.

Check for Reasonableness Compare to your estimate. $\frac{3}{10} \approx \frac{1}{2}$

Practice

Subtract. Write each in simplest form.

1. $\frac{1}{2} - \frac{1}{4} =$ _____

2. $\frac{7}{8} - \frac{1}{4} =$ _____

3. $\frac{7}{12} - \frac{1}{6} =$ _____

Problem Solving

4. The average rainfall in April and October for Springfield is shown in the table below. How much more rain falls on average in April than in October?

Average Rainfall for Springfield	
Month	**Rainfall (in.)**
April	$\frac{11}{16}$
October	$\frac{3}{8}$

_____

5. Trisha helped clean up her neighborhood by picking up plastic. She collected $\frac{3}{4}$ pound of plastic the first day and $\frac{1}{6}$ pound of plastic the second day. How much more trash did she collect the first day than the second day?

_____

6. Wyatt is hiking a trail that is $\frac{11}{12}$ mile long. After hiking $\frac{1}{4}$ mile, he stops for water. How much farther must he hike to finish the trail?

_____

Test Practice

7. The table shows the distance each student ran on Wednesday. How much farther did Joey run than Steve?

ⓐ $\frac{1}{12}$ mile

ⓒ $\frac{1}{2}$ mile

ⓑ $\frac{5}{12}$ mile

ⓓ $\frac{5}{6}$ mile

Student	Distance (mi)
Steve	$\frac{1}{6}$
Charlie	$\frac{1}{4}$
Joey	$\frac{2}{3}$

Problem-Solving Investigation

STRATEGY: Determine Reasonable Answers

Lesson 8

ESSENTIAL QUESTION
How can equivalent fractions help me add and subtract fractions?

Learn the Strategy

Leandra feeds her pet rabbit Bounce the same amount of food each day. Bounce eats three times a day. About how much food does Leandra feed Bounce each day?

Time	Food (cups)
Morning	$\frac{3}{4}$
Afternoon	$\frac{3}{4}$
Evening	$\frac{1}{4}$

 **Understand**

What facts do you know?

Leandra feeds the rabbit the same amount every _____.

What do you need to find?

how much food she feeds her rabbit each _____

 **Plan**

Use estimation to find a reasonable answer.

Supper time?

 **Solve**

Round each fraction to the nearest whole number.

Morning   **Afternoon**   **Evening**

$\frac{3}{4} \longrightarrow$ ___ $\frac{3}{4} \longrightarrow$ ___ $\frac{1}{4} \longrightarrow$ ___

In one day, she feeds Bounce about ___ + ___ + ___, or ___ cups of food.

So, Leandra feeds Bounce about ___ cups of food each day.

Check

Is my answer reasonable?
The estimate of 2 cups is close to the actual amount of $1\frac{3}{4}$ cups of food.

Practice the Strategy

Jonas needs to determine how much wood to buy at the store to make picture frames. The dimensions for each picture frame are listed in the table. About how many feet of wood does Jonas need to buy to build 5 frames?

Frame	Wood (feet)
Top	$\frac{5}{8}$
Bottom	$\frac{5}{8}$
Left Side	$\frac{3}{4}$
Right Side	$\frac{3}{4}$

1 Understand

What facts do you know?

_____

What do you need to find?

_____

2 Plan

_____

3 Solve

4 Check

Is my answer reasonable?

_____

_____

Apply the Strategy

Determine a reasonable answer to solve each problem.

1. Use the table to determine whether 245 pounds, 260 pounds, or 263 pounds is the most reasonable estimate for how much more the ostrich weighs than the flamingo. Explain.

Bird	Weight (lb)
Flamingo	$9\frac{1}{10}$
Ostrich	$253\frac{1}{2}$

My Work!

_____

_____

_____

2. **Mathematical PRACTICE 6 Explain to a Friend** A grocer sells 12 pounds of apples. Of those, $5\frac{3}{4}$ pounds are green and $3\frac{1}{4}$ pounds are golden. The rest are red. Which is a more reasonable estimate for how many pounds of red apples the grocer sold: 3 pounds or 5 pounds? Explain.

_____

_____

3. A puzzle book costs $4.25. A novel costs $0.70 more than the puzzle book. Which is the most reasonable estimate for the total cost of both items: $14, $16, or $18?

_____

4. Thirty students from the Netherlands set up a record of 1,500,000 dominoes. Of these, 1,138,101 were toppled by one push. Which is a more reasonable estimate for how many dominoes remained standing after that push: 350,000 or 400,000?

_____

_____

Review the Strategies

Use any strategy to solve each problem.

- Determine reasonable answers.
- Look for a pattern.
- Solve a simpler problem.
- Find an estimate or exact answer.

5. Use the graph below. Is 21 inches, 24 inches, or 25 inches a more reasonable total amount of rain that fell in May, June, and July?

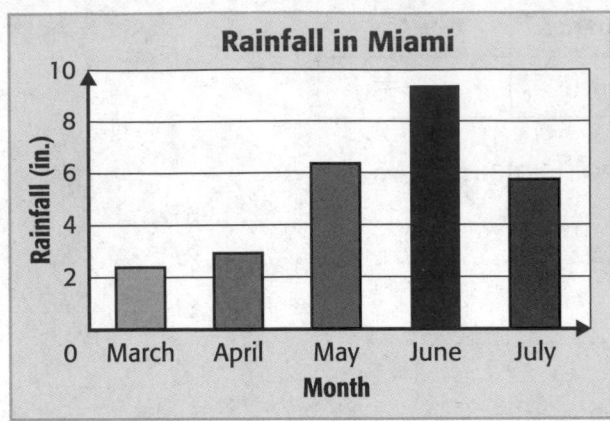

6. **Mathematical PRACTICE** **5** **Use Math Tools** Draw the next figure in the pattern.

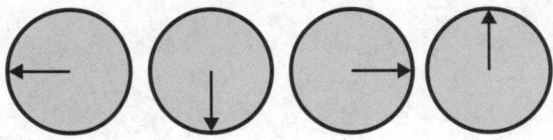

7. A high jumper starts the bar at 48 inches and raises the bar $\frac{1}{2}$ inch after each jump. How high will the bar be after the seventh jump?

8. Ms. Kennedy wants to buy a new interactive whiteboard for her classroom that costs $989. So far, Ms. Kennedy has collected $485 in donations, and $106 in fundraising. About how much more money does she need to buy the interactive whiteboard?

My Work!

MY Homework

Homework Helper

Need help? connectED.mcgraw-hill.com

Ms. Montoya makes $2\frac{3}{4}$ pounds of goat cheese in the morning. In the afternoon, she makes $1\frac{1}{5}$ pounds of goat cheese. Is 3 pounds, 4 pounds, or 5 pounds the most reasonable estimate for how much goat cheese Ms. Montoya makes in one day?

1 Understand

What facts do you know?
Ms. Montoya makes $2\frac{3}{4}$ pounds and $1\frac{1}{5}$ pounds of goat cheese in one day.

What do you need to find?
a reasonable estimate of how much goat cheese she makes in one day

2 Plan

Use estimation to find a reasonable answer.

3 Solve

Round each amount of cheese to the nearest whole number.

Morning   **Afternoon**

$2\frac{3}{4} \rightarrow 3$ $1\frac{1}{5} \rightarrow 1$

In one day, she makes about $3 + 1$, or 4 pounds of goat cheese.

So, Ms. Montoya makes about 4 pounds of goat cheese in one day.

4 Check

Is my answer reasonable?
Subtract the rounded amount she made in the morning from the estimated total.

$4 - 3 = 1$

So, the answer is reasonable.

 **Problem Solving**

1. Alyssa needs $7\frac{5}{8}$ inches of ribbon for one project and $4\frac{7}{8}$ inches of ribbon for another project. If she has 11 inches of ribbon, will she have enough to complete both projects? Explain.

   _____

   _____

2. Josiah has a piece of wood that measures $10\frac{1}{8}$ feet. He wants to make 5 shelves. If each shelf is $1\frac{3}{4}$ feet long, does he have enough to make 5 shelves? Explain.

   _____

   _____

3. After school, Philipe spent $1\frac{3}{4}$ hours at baseball practice, $2\frac{1}{4}$ hours on homework, and $\frac{1}{4}$ hour getting ready for bed. Which is the most reasonable estimate for how long he spent on his activities: 3 hours, 4 hours, or 5 hours? Explain.

   _____

   _____

   _____

4. Kimane and a friend picked apples. Kimane picked $4\frac{2}{3}$ pounds and his friend picked $5\frac{5}{6}$ pounds. Which is the most reasonable estimate for how many pounds they picked altogether: 9 pounds, 11 pounds, or 12 pounds?

   _____

   _____

Estimate Sums and Differences

 **Math in My World** Watch Tutor

Scuba–DUBA–Doo

Example 1

Lucita and Alexis went scuba diving for $7\frac{1}{3}$ hours. The next day, they went scuba diving for $4\frac{2}{3}$ hours. About how many hours did they spend scuba diving altogether?

Estimate $7\frac{1}{3} + 4\frac{2}{3}$.

Round each mixed number to the nearest whole number.

$\frac{1}{3}$ is less than $\frac{1}{2}$.
So, $7\frac{1}{3}$ rounds down to 7.

$\frac{2}{3}$ is greater than $\frac{1}{2}$.
So, $4\frac{2}{3}$ rounds up to 5.

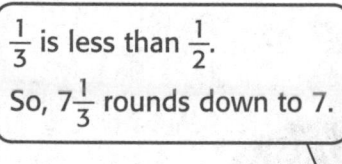

$$7\frac{1}{3} + 4\frac{2}{3} \approx \rule{2cm}{0.4pt} + \rule{2cm}{0.4pt}$$

$$\approx \rule{2cm}{0.4pt}$$

Helpful Hint
The ≈ means *about* or *approximately equal to.*

So, $7\frac{1}{3} + 4\frac{2}{3} \approx \rule{2cm}{0.4pt}$.

Lucita and Alexis spent about \rule{2cm}{0.4pt} hours scuba diving altogether.

Example 2

One sea anemone is $3\frac{1}{4}$ feet across while a second sea anemone is $5\frac{3}{4}$ feet across. About how much longer across is the second anemone?

Estimate $5\frac{3}{4} - 3\frac{1}{4}$.

Round each mixed number to the nearest whole number.

| $\frac{3}{4}$ is greater than $\frac{1}{2}$. So, $5\frac{3}{4}$ rounds up to 6. | $\frac{1}{4}$ is less than $\frac{1}{2}$. So, $3\frac{1}{4}$ rounds down to 3. |

$$5\frac{3}{4} - 3\frac{1}{4} \approx \underline{\hspace{2cm}} - \underline{\hspace{2cm}}$$

$$\approx \underline{\hspace{2cm}}$$

So, $5\frac{3}{4} - 3\frac{1}{4} \approx \underline{\hspace{2cm}}$.

The second sea anemone is about _____ feet longer across than the first sea anemone.

Guided Practice

Estimate by rounding each mixed number to the nearest whole number.

1. $2\frac{1}{8} + 3\frac{5}{8}$

_____

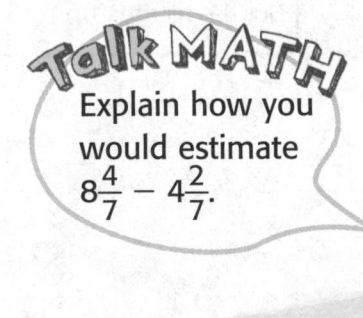
Talk MATH
Explain how you would estimate $8\frac{4}{7} - 4\frac{2}{7}$.

2. $5\frac{2}{9} - 3\frac{2}{3}$

_____

3. $3\frac{3}{5} + 1\frac{1}{5}$

_____

Independent Practice

Estimate by rounding each mixed number to the nearest whole number.

4. $7\frac{2}{3}$
$-\ 4\frac{1}{4}$
_____

5. $5\frac{1}{3}$
$+\ 3\frac{7}{9}$
_____

6. $9\frac{1}{7}$
$-\ 5\frac{6}{7}$
_____

7. $6\frac{7}{10} - 1\frac{1}{5}$

_____

8. $8\frac{11}{12} + 4\frac{1}{3}$

_____

9. $15\frac{3}{7} - 3\frac{4}{7}$

_____

10. $10\frac{2}{7}$
$+\ 7\frac{5}{7}$
_____

11. $13\frac{4}{11}$
$-\ 4\frac{1}{4}$
_____

12. $12\frac{7}{10}$
$+\ 9\frac{3}{5}$
_____

13. $19\frac{3}{7} + \frac{13}{14}$

_____

14. $7\frac{7}{9} - 1\frac{5}{18}$

_____

15. $\frac{9}{16} + 16\frac{5}{8}$

_____

Problem Solving

16. Liam spent $1\frac{3}{4}$ hours playing board games and $2\frac{1}{4}$ hours watching a movie. About how much time did Liam spend altogether on these two activities? Round each mixed number to the nearest whole number.

17. **Mathematical PRACTICE 2 Use Number Sense** Beth walked $10\frac{7}{8}$ miles one week. She walked $2\frac{1}{4}$ fewer miles the following week. About how many miles did she walk the second week? Round each mixed number to the nearest whole number.

18. Patrick has played guitar for $3\frac{5}{6}$ years. Gen has played guitar for $6\frac{1}{12}$ years. Estimate how many more years Gen has played guitar than Patrick. Round each mixed number to the nearest whole number.

My Work!

HOT Problems

19. **Mathematical PRACTICE 3 Justify Conclusions** Select two mixed numbers whose estimated difference is 1. Justify your selection.

20. **Building on the Essential Question** How does number sense of fractions help when estimating sums and differences?

MY Homework

Homework Helper

Need help? connectED.mcgraw-hill.com

Estimate $3\frac{5}{8} - 1\frac{1}{8}$.

Round each mixed number to the nearest whole number.

$\frac{5}{8}$ is greater than $\frac{1}{2}$.

So, $3\frac{5}{8}$ rounds up to 4.

$\frac{1}{8}$ is less than $\frac{1}{2}$.

So, $1\frac{1}{8}$ rounds down to 1.

$$3\frac{5}{8} - 1\frac{1}{8} \approx 4 - 1$$
$$\approx 3$$

So, $3\frac{5}{8} - 1\frac{1}{8}$ is about 3.

Practice

Estimate by rounding each mixed number to the nearest whole number.

1. $2\frac{1}{5} + 6\frac{4}{5}$

2. $6\frac{5}{12} - 2\frac{1}{12}$

3. $11\frac{8}{9} + 4\frac{1}{3}$

4. $7\frac{1}{6} - 5\frac{3}{4}$

5. $\frac{11}{18} + 6\frac{1}{10}$

6. $15\frac{1}{3} - 2\frac{5}{8}$

Problem Solving

For Exercises 7–9, use the pictures shown.

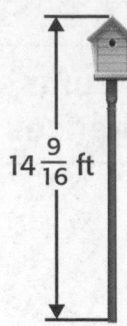

$14\frac{9}{16}$ ft

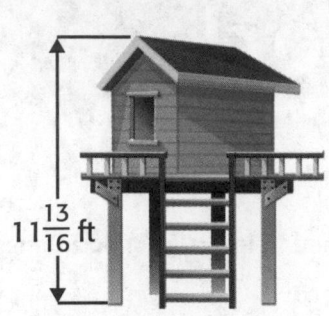

$11\frac{13}{16}$ ft

$8\frac{1}{4}$ ft

7. About how much taller is the birdhouse than the swing set? Round each mixed number to the nearest whole number.

_____

8. About how much taller is the birdhouse than the tree house? Round each mixed number to the nearest whole number.

_____

9. Mathematical **PRACTICE 1** **Plan Your Solution** Find the approximate height difference between the birdhouse and the tree house. Is it greater than or less than the approximate difference in height between the swing set and the tree house? Round each mixed number to the nearest whole number.

_____

_____

My Work!

Test Practice

10. Use the table at the right. About how many hours did Frank ride his bike altogether?

Ⓐ 4 hours   Ⓒ 6 hours

Ⓑ 5 hours   Ⓓ 7 hours

Biking Time	
Week	Length (h)
1	$3\frac{1}{4}$
2	$2\frac{5}{6}$

Check My Progress

Vocabulary Check

Use the word bank below to complete each sentence.

benchmark fractions   **denominator**   **least common denominator (LCD)**
mixed number   **numerator**   **unknown**

1. The _____denominator_____ is the bottom number in a fraction.

2. A number that has a whole number part and a fraction part is a
_____mixed number_____.

3. A missing value in a number sentence or equation is the
_____unknown_____.

4. The _____numerator_____ is the top number in a fraction.

Concept Check

Subtract. Write the difference in simplest form.

5. $\frac{7}{10} - \frac{1}{2} =$ _____$\frac{1}{5}$_____

6. $\frac{5}{8} - \frac{1}{6} =$ _____$\frac{4}{8}$_____

7. $\frac{7}{12} - \frac{1}{3} =$ _____$\frac{6}{3}$_____

Estimate by rounding each mixed number to the nearest whole number.

8. $8\frac{9}{15} - \frac{4}{5}$ _____$8\frac{5}{10}$_____

9. $\frac{9}{16} + 16\frac{5}{8}$ _____$16-$_____

10. $5\frac{6}{11} - 3\frac{4}{5}$ _____

Problem Solving

11. The table shows how many hours Allie swam in two weeks. For about how many hours did she swim altogether? Round each mixed number to the nearest whole number.

Day	Length (h)
Week 1	$4\frac{1}{3}$
Week 2	$2\frac{5}{6}$

_____

12. To clean a large painting, you need $3\frac{1}{4}$ ounces of cleaner. A small painting requires only $2\frac{3}{4}$ ounces. You have 8 ounces of cleaner. About how many ounces of cleaner will you have left if you clean a large and a small painting? Round each mixed number to the nearest whole number.

_____

13. Andrew plays football. On one play, he ran the ball $24\frac{1}{3}$ yards. The following play, he was tackled and lost $3\frac{2}{3}$ yards. The next play, he ran for $5\frac{1}{4}$ yards. Estimate how much farther the ball is down the field after these three plays. Round each mixed number to the nearest whole number.

_____

14. Marsha and Tegene are making table decorations for a party. Marsha made $\frac{2}{9}$ of a decoration in half an hour. Tegene made $\frac{2}{3}$ of a decoration in the same amount of time. How much more of a decoration did Tegene make in half an hour?

_____

Test Practice

15. Greta's family drinks $3\frac{1}{4}$ quarts of milk and $1\frac{5}{6}$ quarts of juice in one week. About how much more milk than juice does Greta's family drink in one week?

Ⓐ $\frac{1}{2}$ quart   Ⓒ 2 quarts

Ⓑ 1 quart   Ⓓ 3 quarts

Hands On
Use Models to Add Mixed Numbers

Lesson 10

ESSENTIAL QUESTION
How can equivalent fractions help me add and subtract fractions?

You can use fraction circles to add mixed numbers.

Draw It

Find $2\frac{1}{3} + 1\frac{1}{3}$.

1. Shade and label the fraction circles to represent each mixed number.

$2\frac{1}{3}$ →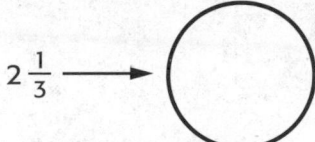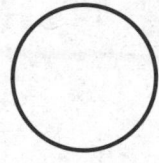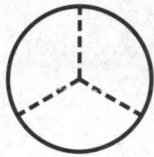

$1\frac{1}{3}$ →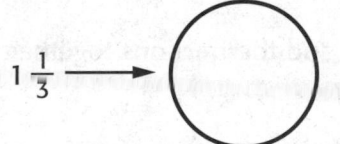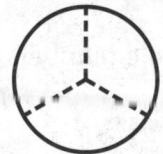

2. Combine the whole numbers and fractions.

How many whole fraction circles are there altogether? _____

How many thirds are there altogether? _____

Add the whole numbers and the fractions.

$$2\frac{1}{3} + 1\frac{1}{3} = 1 + 1 + \frac{1}{3} + 1 + \frac{1}{3}$$

$$= 1 + 1 + 1 + \frac{1}{3} + \frac{1}{3} \quad \text{Group the whole numbers and fractions together.}$$

$$= 3 + \frac{2}{3}$$

$$= 3\frac{2}{3}$$

So, $2\frac{1}{3} + 1\frac{1}{3} = \boxed{}\frac{\boxed{}}{\boxed{}}$.

Try It

Find $1\frac{7}{8} + 1\frac{1}{2}$.

1 Shade and label the fraction circles to represent each mixed number. Write $1\frac{1}{2}$ as the equivalent fraction $1\frac{4}{8}$.

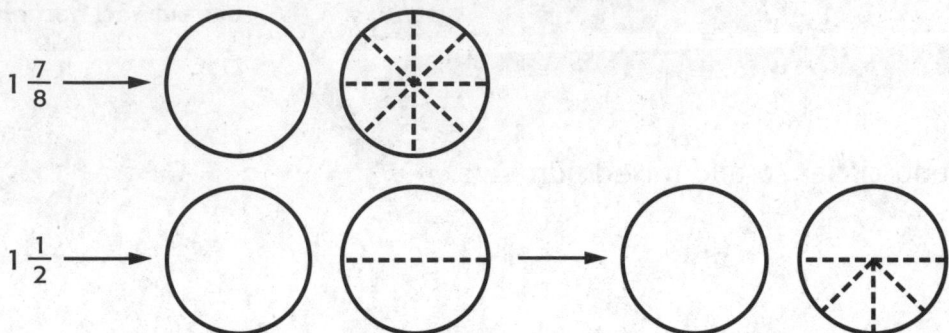

$1\frac{7}{8} \longrightarrow$

$1\frac{1}{2} \longrightarrow$

2 Combine the whole numbers and fractions.

How many whole fraction circles are there altogether? _____

How many eighths are there altogether? _____

Add.

$$1\frac{7}{8} + 1\frac{4}{8} = 1 + \frac{7}{8} + 1 + \frac{4}{8}$$

$$= 1 + 1 + \frac{7}{8} + \frac{4}{8} \qquad \text{Group the whole numbers and the fractions together.}$$

$$= 2 + \frac{11}{8}$$

$$= 2 + 1\frac{3}{8} \qquad \text{Write } \frac{11}{8} \text{ as } \frac{8}{8} + \frac{3}{8}, \text{ or } 1\frac{3}{8}.$$

$$= 3\frac{3}{8}$$

So, $1\frac{7}{8} + 1\frac{1}{2} = \boxed{} \frac{\boxed{}}{\boxed{}}$.

Talk About It

1. **Mathematical PRACTICE ③ Draw a Conclusion** Refer to Step 1 of the above activity. Explain why it was necessary to write $1\frac{1}{2}$ as $1\frac{4}{8}$.

_____

_____

Practice It

Shade the fraction circles to represent each mixed number. Then find each sum.

2. $1\frac{1}{5} + 2\frac{3}{5} =$ _____

$1\frac{1}{5} \rightarrow$ ◯ ⊛

$2\frac{3}{5} \rightarrow$ ◯ ◯ ⊛

3. $2\frac{5}{6} + 1\frac{1}{6} =$ _____

$2\frac{5}{6} \rightarrow$ ◯ ◯ ⊛

$1\frac{1}{6} \rightarrow$ ◯ ⊛

4. $2\frac{2}{3} + 1\frac{1}{2} =$ _____

$2\frac{2}{3} \rightarrow$ ◯ ◯ ⊛

$1\frac{1}{2} \rightarrow$ ◯ ⊛

5. $2\frac{7}{8} + 1\frac{1}{4} =$ _____

$2\frac{7}{8} \rightarrow$ ◯ ◯ ⊛

$1\frac{1}{4} \rightarrow$ ◯ ⊛

6. $2\frac{1}{2} + 2\frac{1}{4} =$ _____

$2\frac{1}{2} \rightarrow$ ◯ ◯ ⊕

$2\frac{1}{4} \rightarrow$ ◯ ◯ ⊕

7. $2\frac{3}{4} + 2\frac{1}{8} =$ _____

$2\frac{3}{4} \rightarrow$ ◯ ◯ ⊛

$2\frac{1}{8} \rightarrow$ ◯ ◯ ⊛

 **Apply It**

8. Eduardo worked $1\frac{1}{2}$ hours on Monday, $1\frac{1}{2}$ hours on Tuesday, and $2\frac{1}{2}$ hours on Wednesday. How many hours did he work in all?

_____

9. Marissa walked $1\frac{1}{4}$ miles to the store. She walked the same distance back home. How far did Marissa walk altogether?

_____

Mathematical
10. **PRACTICE** 4 **Model Math** Write and solve a real-world problem that could be represented by the fraction circles shown.

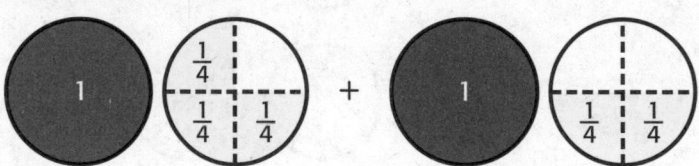

_____

_____

_____

Write About It

11. How can I use fractions circles to find the sum of mixed numbers?

_____

_____

_____

MY Homework

Homework Helper

Need help? connectED.mcgraw-hill.com

Find $1\frac{3}{5} + 1\frac{4}{5}$.

1 Shade and label the fraction circles to represent each mixed number.

$1\frac{3}{5} \rightarrow$ [circle labeled 1] [circle with fifths labeled $\frac{1}{5}$, $\frac{1}{5}$, $\frac{1}{5}$] $1\frac{4}{5} \rightarrow$ [circle labeled 1] [circle with fifths labeled $\frac{1}{5}$, $\frac{1}{5}$, $\frac{1}{5}$, $\frac{1}{5}$]

2 Combine the whole numbers and fractions.

Add.

$$1\frac{3}{5} + 1\frac{4}{5} = 1 + \frac{3}{5} + 1 + \frac{4}{5}$$

$$= 1 + 1 + \frac{3}{5} + \frac{4}{5} \quad \text{Group the whole numbers and the fractions together.}$$

$$= 2 + \frac{7}{5}$$

$$= 2 + 1\frac{2}{5} \quad \text{Write } \frac{7}{5} \text{ as } \frac{5}{5} + \frac{2}{5}, \text{ or } 1\frac{2}{5}.$$

$$= 3\frac{2}{5}$$

So, $1\frac{3}{5} + 1\frac{4}{5} = 3\frac{2}{5}$.

Practice

1. Shade the fraction circles to represent each mixed number. Then find the sum.

$1\frac{1}{2} + 1\frac{2}{3} =$ _____ $1\frac{1}{2} \rightarrow$  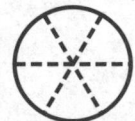 $1\frac{2}{3} \rightarrow$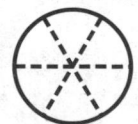

2. Shade the fraction circles to represent each mixed number. Then find the sum.

$1\frac{3}{5} + 2\frac{2}{5} =$ _____ $1\frac{3}{5} \rightarrow$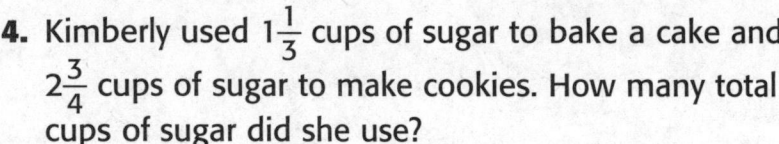

$2\frac{2}{5} \rightarrow$

 # Problem Solving

3. Andrew spent $1\frac{1}{6}$ hours studying for his science exam. He spent another $1\frac{1}{2}$ hours studying for his history exam. How many total hours did Andrew spend studying for his exams? Draw fraction circles to solve.

_____

_____

4. Kimberly used $1\frac{1}{3}$ cups of sugar to bake a cake and $2\frac{3}{4}$ cups of sugar to make cookies. How many total cups of sugar did she use?

_____

_____

My Drawing!

5. Mathematical
PRACTICE **3** **Which One Doesn't Belong?** Find each sum by drawing fraction circles. Then circle the expression that does not belong. Explain.

$1\frac{1}{2} + 2\frac{1}{3}$	$2\frac{2}{3} + 1\frac{1}{6}$
$1\frac{5}{6} + 2\frac{1}{2}$	$1\frac{2}{3} + 2\frac{1}{6}$

_____

_____

Add Mixed Numbers

Lesson 11

ESSENTIAL QUESTION
How can equivalent fractions help me add and subtract fractions?

 ## Math in My World

Example 1

A hammerhead shark swam $2\frac{1}{4}$ miles. The next day, it swam $1\frac{1}{4}$ miles. How many miles did it swim altogether?

Find $2\frac{1}{4} + 1\frac{1}{4}$.

Estimate $2\frac{1}{4} + 1\frac{1}{4} \approx 2 + 1$, or 3

$2\frac{1}{4} + 1\frac{1}{4} = 1 + 1 + \frac{1}{4} + 1 + \frac{1}{4}$ Write as a sum of wholes and fractions.

$= 1 + 1 + 1 + \frac{1}{4} + \frac{1}{4}$ Group the wholes and the fractions together.

$= 3 + \frac{2}{4}$ $1 + 1 + 1 = 3$ and $\frac{1}{4} + \frac{1}{4} = \frac{2}{4}$

$= \boxed{}\ \frac{\boxed{}}{\boxed{}}$ Write in simplest form. $\frac{2}{4} = \frac{1}{2}$

So, the hammerhead shark swam $\boxed{}\ \frac{\boxed{}}{\boxed{}}$ miles.

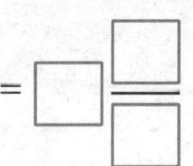

Let's hammer out the answer!

The models show that $2\frac{1}{4} + 1\frac{1}{4} = 3\frac{1}{2}$.

Check Compared to the estimate, $3\frac{1}{2} \approx 3$. The answer is reasonable.

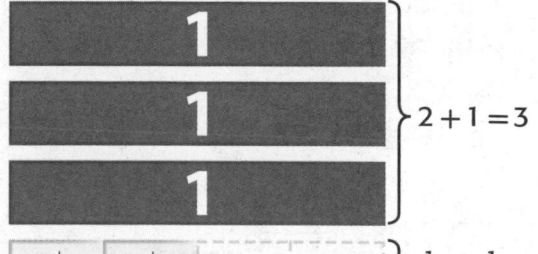

Example 2

The diagram shows the length of a sea turtle. What is the total length of the sea turtle?

Find $\frac{7}{8} + 3\frac{1}{4} + 1\frac{1}{8}$.

$\frac{7}{8}$ ft $3\frac{1}{4}$ ft $1\frac{1}{8}$ ft

1 Write an equivalent fraction for $3\frac{1}{4}$ so that the fractions all have the same denominator. The LCD is 8.

$$3\frac{1}{4} = 3\frac{1 \times \boxed{2}}{4 \times \boxed{2}} = 3\frac{2}{8}$$

Write an equivalent fraction with a denominator of 8.

2 Add.

$$\frac{7}{8} + 3\frac{1}{4} + 1\frac{1}{8} = \frac{7}{8} + 3\frac{2}{8} + 1\frac{1}{8}$$ Write $3\frac{1}{4}$ as $3\frac{2}{8}$.

$$= 3 + 1 + \frac{7}{8} + \frac{2}{8} + \frac{1}{8}$$ Group the wholes and the fractions together.

$$= 4 + \frac{10}{8}$$ $3 + 1 = 4$ and $\frac{7}{8} + \frac{2}{8} + \frac{1}{8} = \frac{10}{8}$

$$= 4 + 1\frac{2}{8}$$ Write $\frac{10}{8}$ as $\frac{8}{8} + \frac{2}{8}$, or $1\frac{2}{8}$.

$$= \boxed{}\frac{\boxed{}}{\boxed{}}$$ Write in simplest form. $4 + 1 = 5$ and $\frac{2}{8} = \frac{1}{4}$

So, $\frac{7}{8} + 3\frac{1}{4} + 1\frac{1}{8} = \boxed{}\frac{\boxed{}}{\boxed{}}$.

The total length of the sea turtle is $\boxed{}\frac{\boxed{}}{\boxed{}}$ feet.

Guided Practice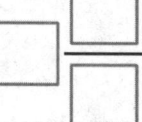

Talk MATH

Explain how to simplify $3\frac{6}{4}$.

1. Estimate, then add. Write the sum in simplest form.

$$3\frac{3}{8} + 2\frac{1}{2} = 1 + 1 + 1 + \frac{3}{8} + 1 + 1 + \frac{4}{8}$$

$$= \boxed{}\frac{\boxed{}}{\boxed{}}$$

678 Chapter 9 Add and Subtract Fractions

Name _____

Independent Practice

Estimate, then add. Write each sum in simplest form.

2. $4\frac{3}{5} + 3\frac{1}{5} =$ _____

3. $7\frac{4}{11} + 2\frac{6}{11} =$ _____

4. $5\frac{1}{12} + 6\frac{1}{4} =$ _____

5. $8\frac{4}{15} + 3\frac{2}{15} =$ _____

6. $6\frac{1}{9} + 2\frac{1}{3} =$ _____

7. $5\frac{1}{3} + 6\frac{1}{2} =$ _____

8. $\begin{array}{r} 3\frac{4}{9} \\ + 4\frac{2}{3} \\ \hline \end{array}$

9. $\begin{array}{r} 6\frac{3}{4} \\ + 3\frac{1}{8} \\ \hline \end{array}$

10. $\begin{array}{r} 4\frac{3}{7} \\ + 7\frac{1}{2} \\ \hline \end{array}$

Algebra Find each unknown.

11. $9\frac{9}{10} + 7\frac{3}{5} = y$

12. $14\frac{19}{20} + 8\frac{1}{4} = k$

13. $16\frac{11}{12} + 5\frac{2}{3} = d$

$y =$ _____

$k =$ _____

$d =$ _____

Problem Solving

14. Zita made $1\frac{5}{8}$ quarts of punch. Then she made $1\frac{7}{8}$ more quarts. How much punch did she make in all?

_____

15. Find *five and two-eighths* plus *three and six-eighths*. Write in words in simplest form.

_____

16. Mathematical **PRACTICE** ➊ **Make a Plan**
Tomás made fruit salad using the recipe. How many cups of fruit are needed altogether?

Recipe for Fruit Salad	
$3\frac{1}{2}$ c	Bananas
$1\frac{1}{4}$ c	Grapes
$1\frac{3}{4}$ c	Strawberries
$2\frac{1}{4}$ c	Pears

_____

HOT Problems

17. Mathematical **PRACTICE** ➌ **Find the Error** Antwan is finding $4\frac{1}{5} + 2\frac{3}{5}$. Find and correct his mistake.

$$4\frac{1}{5} + 2\frac{3}{5} = 6\frac{4}{10}$$

_____

18. ❓ **Building on the Essential Question** How can equivalent fractions help when adding mixed numbers?

_____

_____

_____

MY Homework

Homework Helper eHelp

Need help? connectED.mcgraw-hill.com

Find $4\frac{3}{8} + 7\frac{1}{4}$.

 Write an equivalent fraction for $7\frac{1}{4}$ so that the fractions all have the same denominators. The LCD is 8.

$$7\frac{1}{4} = 7\frac{1 \times 2}{4 \times 2} = 7\frac{2}{8}$$ Write an equivalent fraction with a denominator of 8.

 Add.

$$4\frac{3}{8} + 7\frac{1}{4} = 4\frac{3}{8} + 7\frac{2}{8}$$ Write $7\frac{1}{4}$ as $7\frac{2}{8}$.

$$= 4 + 7 + \frac{3}{8} + \frac{2}{8}$$ Group the wholes and the fractions together.

$$= 11\frac{5}{8}$$

So, $4\frac{3}{8} + 7\frac{1}{4} = 11\frac{5}{8}$.

Practice

Estimate, then add. Write each sum in simplest form.

1. $2\frac{1}{10} + 5\frac{7}{10} =$ _____

2. $9\frac{3}{4} + 8\frac{3}{4} =$ _____

3. $3\frac{5}{8} + 6\frac{1}{2} =$ _____

4. $1\frac{1}{12} + 4\frac{5}{12} =$ _____

5. $11\frac{3}{5} + 6\frac{4}{15} =$ _____

6. $9\frac{1}{2} + 12\frac{11}{20} =$ _____

Problem Solving

7. A flower is $9\frac{3}{4}$ inches tall. In one week, it grew $1\frac{1}{8}$ inches. How tall is the flower at the end of the week? Write in simplest form.

_____

8. Find _ten and three-sevenths_ plus _eighteen and two-sevenths._ Write in words in simplest form.

_____

9. Mathematical **PRACTICE** 6 **Explain to a Friend** Connor is filling a 15-gallon wading pool. On his first trip, he carried $3\frac{1}{12}$ gallons of water. He carried $3\frac{5}{6}$ gallons on his second trip and $3\frac{1}{2}$ gallons on his third trip. Suppose he carries 5 gallons on his next trip. Will the pool be filled? Explain.

_____

_____

Test Practice

10. Benjamin had $2\frac{1}{3}$ gallons of fruit punch left after a party. He had $1\frac{3}{4}$ gallons of lemonade left. How many total gallons did he have left?

Ⓐ $4\frac{1}{12}$ gallons

Ⓒ $1\frac{5}{12}$ gallons

Ⓑ $3\frac{1}{12}$ gallons

Ⓓ $\frac{5}{12}$ gallon

Number and Operations – Fractions
5.NF.1, 5.NF.2

Subtract Mixed Numbers

Lesson 12

ESSENTIAL QUESTION
How can equivalent fractions help me add and subtract fractions?

 **Math in My World**  Watch  Tools  Tutor

Example 1

One King crab weighs $2\frac{3}{4}$ pounds. A second King crab weighs $1\frac{1}{4}$ pounds. How much more does the one King crab weigh? Use models to find the difference.

Find $2\frac{3}{4} - 1\frac{1}{4}$.

Estimate $3 - 1 =$ _____

Model $2\frac{3}{4}$ using fraction tiles.

Subtract $1\frac{1}{4}$ by crossing out 1 whole and one $\frac{1}{4}$-tile.

There is one whole and two $\frac{1}{4}$-tiles left,

which is $1\frac{2}{4}$, or ☐ $\dfrac{☐}{☐}$.

So, $2\frac{3}{4} - 1\frac{1}{4} =$ ☐ $\dfrac{☐}{☐}$.

The first King crab weighs ☐ $\dfrac{☐}{☐}$ pounds more than the second.

Check for Reasonableness _____ $\approx$ ☐ $\dfrac{☐}{☐}$

1
1

| $\frac{1}{4}$ | $\frac{1}{4}$ | $\frac{1}{4}$ | |

Example 2

Find $6\frac{11}{16} - 2\frac{5}{8}$.

Estimate $7 - 3 =$ _____

1 Write an equivalent fraction for $2\frac{5}{8}$ so that the fractions have the same denominator. The LCD is 16.

$$2\frac{5}{8} = 2\frac{5 \times \boxed{2}}{8 \times \boxed{2}} \rightarrow \boxed{}\,\frac{\boxed{}}{\boxed{}}$$

2 Subtract the wholes.
Then subtract the fractions.

$$6\frac{11}{16} - 2\frac{5}{8}$$

$\downarrow \qquad\qquad \downarrow$

Subtract the wholes.
$6 - 2 = 4$
Subtract the fractions.
$\frac{11}{16} - \frac{10}{16} = \frac{1}{16}$

$\rightarrow \quad 6\frac{11}{16} - 2\frac{\boxed{}}{\boxed{}} = \boxed{}\,\frac{\boxed{}}{\boxed{}}$

So, $6\frac{11}{16} - 2\frac{5}{8} = \boxed{}\,\frac{\boxed{}}{\boxed{}}$.

Check for Reasonableness _____ $\approx \boxed{}\,\frac{\boxed{}}{\boxed{}}$

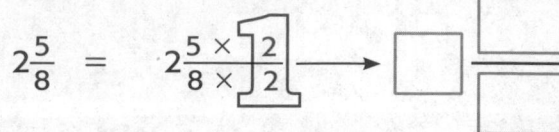

Talk MATH

Describe the steps you would take to find $3\frac{5}{8} - 2\frac{3}{8}$.

Guided Practice ✓ Check

Estimate, then subtract. Write each difference in simplest form.

1.
$$\begin{array}{r} 4\frac{2}{3} \\ - 2\frac{1}{3} \\ \hline \end{array}$$
$\boxed{}\,\frac{\boxed{}}{\boxed{}}$

2.
$$\begin{array}{r} 5\frac{4}{5} \\ - 3\frac{2}{5} \\ \hline \end{array}$$
$\boxed{}\,\frac{\boxed{}}{\boxed{}}$

Independent Practice

Estimate, then subtract. Write each difference in simplest form.

3.
$$5\frac{3}{4}$$
$$-\ 2\frac{1}{2}$$

4.
$$6\frac{5}{7}$$
$$-\ 3\frac{3}{7}$$

5.
$$7\frac{8}{9}$$
$$-\ 5\frac{1}{3}$$

6.
$$15\frac{11}{12}$$
$$-\ 4\frac{1}{3}$$

7.
$$13\frac{9}{10}$$
$$-\ 4\frac{2}{5}$$

8.
$$12\frac{5}{6}$$
$$-\ 7\frac{1}{3}$$

9. $8\frac{3}{8} - 2\frac{1}{4} =$ _____

10. $7\frac{7}{8} - 4\frac{1}{2} =$ _____

11. $12\frac{7}{10} - 7\frac{2}{5} =$ _____

Mathematical PRACTICE 2 Use Algebra Find each unknown.

12. $11\frac{11}{12} - 2\frac{1}{12} = x$

13. $14\frac{9}{14} - 5\frac{2}{7} = c$

14. $18\frac{11}{15} - 9\frac{2}{5} = n$

$x =$ _____

$c =$ _____

$n =$ _____

Problem Solving

15. The length of Mr. Cho's garden is $8\frac{5}{6}$ feet. Find the width of Mr. Cho's garden if it is $3\frac{1}{6}$ feet less than the length.

_____

16. Timberly spent $3\frac{4}{5}$ hours and Misty spent $2\frac{1}{10}$ hours at gymnastics practice over the weekend. How many more hours did Timberly spend than Misty at gymnastics practice?

_____

17. Mathematical **PRACTICE** **1** **Make Sense of Problems** Warner lives $9\frac{1}{4}$ blocks away from the ocean. Shelly lives $12\frac{7}{8}$ blocks away from the ocean. How many more blocks does Shelly live away from the ocean than Warner?

My Work!

HOT Problems

18. Mathematical **PRACTICE** **4** **Model Math** Write a real-world problem involving the subtraction of two mixed numbers whose difference is less than $2\frac{1}{2}$. Then solve.

_____

_____

_____

19. **Building on the Essential Question** How can number sense help to know if I have subtracted two mixed numbers correctly?

_____

_____

_____

MY Homework

Homework Helper

Need help? connectED.mcgraw-hill.com

Find $8\frac{1}{2} - 3\frac{1}{6}$.

Estimate $9 - 3 = 6$

1 Write an equivalent fraction for $8\frac{1}{2}$ so that the fractions have the same denominator. The LCD is 6.

$$8\frac{1}{2} = 8\frac{1 \times 3}{2 \times 3} \rightarrow 8\frac{3}{6}$$

2 Subtract the wholes. Then subtract the fractions.

$$8\frac{1}{2} \rightarrow 8\frac{3}{6}$$
$$- 3\frac{1}{6} \rightarrow - 3\frac{1}{6}$$
$$\overline{} \rightarrow 5\frac{2}{6} \text{ or } 5\frac{1}{3}$$

> Subtract the wholes.
> $8 - 3 = 5$
> Subtract the fractions.
> $\frac{3}{6} - \frac{1}{6} = \frac{2}{6}$ or $\frac{1}{3}$

So, $8\frac{1}{2} - 3\frac{1}{6} = 5\frac{2}{6}$ or $5\frac{1}{3}$.

Check for Reasonableness $6 \approx 5\frac{1}{3}$

Practice

Estimate, then subtract. Write each difference in simplest form.

1. $6\frac{5}{8}$
   $- 2\frac{3}{8}$

2. $9\frac{3}{4}$
   $- 1\frac{1}{3}$

3. $4\frac{5}{6}$
   $- 4\frac{1}{3}$

Problem Solving

4. Mrs. Gabel bought $7\frac{5}{6}$ gallons of punch for the class party. The students drank $4\frac{1}{2}$ gallons of punch. How much punch was left at the end of the party? Write in simplest form.

_____

5. Bella is $10\frac{5}{12}$ years old. Franco is $12\frac{7}{12}$ years old. What is the difference in their ages? Write in simplest form.

_____

6. In one week, the fifth grade class recycled $9\frac{2}{3}$ pounds of glass and $12\frac{3}{4}$ pounds of newspaper. How many more pounds of newspaper than glass did the class recycle?

_____

7. **Mathematical PRACTICE** 2 **Use Number Sense** A snack mix recipe calls for $5\frac{3}{4}$ cups of cereal and $3\frac{5}{12}$ cups less of raisins. How many cups of raisins are needed? Write in simplest form.

_____

Test Practice

8. What is the difference between the two weights?

   Ⓐ $\frac{1}{2}$ ounce   Ⓒ $1\frac{3}{8}$ ounces

   Ⓑ $\frac{7}{8}$ ounce   Ⓓ $1\frac{7}{8}$ ounces

$4\frac{1}{2}$ oz  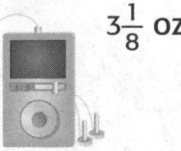 $3\frac{1}{8}$ oz

Subtract with Renaming

Lesson 13

ESSENTIAL QUESTION
How can equivalent fractions help me add and subtract fractions?

Sometimes the fraction in the first mixed number is less than the fraction in the second mixed number. In this case, rename the first mixed number.

 **Math in My World** Watch Tools Tutor

Example 1

A black sea cucumber has an average length of 2 feet. A spotted sea cucumber has an average length of $1\frac{1}{3}$ feet. How much longer is the average black sea cucumber than the average spotted sea cucumber?

Find $2 - 1\frac{1}{3}$.

Estimate $2 - 1 = $ _____

You cannot subtract $\frac{1}{3}$ from 0 thirds.

Rename 2 as $1\frac{3}{3}$ to show more thirds.

$$2 \rightarrow 1\frac{3}{3}$$
$$-1\frac{1}{3} \rightarrow -1\frac{1}{3}$$

Rename 2 as $1\frac{3}{3}$.

Subtract the wholes.
$1 - 1 = 0$
Subtract the fractions.
$\frac{3}{3} - \frac{1}{3} = \frac{2}{3}$

$$2 \quad = \quad 1\frac{3}{3}$$

The average black sea cucumber is $\dfrac{}{}$ foot longer than the average spotted sea cucumber.

Check for Reasonableness $1 \approx \frac{2}{3}$

Online Content at connectED.mcgraw-hill.com

Example 2

Find $4\frac{1}{4} - 2\frac{5}{8}$.

Write equivalent fractions. The LCD is 8.

$$4\frac{1}{4} - 2\frac{5}{8} = 4\frac{1 \times 2}{4 \times 2} - 2\frac{5}{8}$$

$$= 4\frac{2}{8} - 2\frac{5}{8}$$

Helpful Hint
You can always check by using fraction tiles or drawing a picture.

You cannot subtract $\frac{5}{8}$ from $\frac{2}{8}$. So, rename $4\frac{2}{8}$ to show more eighths.

$4\frac{2}{8} \rightarrow \dfrac{\square\ \square}{\square}$ Rename $4\frac{2}{8}$ as $3\frac{10}{8}$.

$-\,2\frac{5}{8} \rightarrow -\,2\frac{5}{8}$

$\dfrac{\square}{\square}$ Subtract.

So, $4\frac{1}{4} - 2\frac{5}{8} = \dfrac{\square\ \square}{\square}$.

Guided Practice

1. Estimate, then subtract. Write each difference in simplest form.

$5\frac{2}{5} \rightarrow \dfrac{\square\ \square}{\square}$

$-\,3\frac{4}{5} \rightarrow -\,3\frac{4}{5}$

$\dfrac{\square}{\square}$

Talk MATH
Describe the steps you would use to find $3\frac{2}{7} - 1\frac{4}{7}$.

Independent Practice

Estimate, then subtract. Write each difference in simplest form.

2. $4\dfrac{3}{8}$
     $-\ 1\dfrac{5}{8}$
     _____

3. $3\dfrac{1}{6}$
     $-\ 1\dfrac{1}{3}$
     _____

4. $5\dfrac{1}{4}$
     $-\ 4\dfrac{1}{2}$
     _____

5. $7\dfrac{1}{2}$
     $-\ 3\dfrac{4}{5}$
     _____

6. 4
     $-\ 1\dfrac{1}{8}$
     _____

7. 12
     $-\ 5\dfrac{1}{6}$
     _____

8. $7\dfrac{2}{7} - 6\dfrac{4}{7} =$ _____

9. $9\dfrac{3}{10} - 5\dfrac{7}{10} =$ _____

10. $10\dfrac{1}{3} - 3\dfrac{2}{3} =$ _____

11. $18 - 9\dfrac{1}{4} =$ _____

12. $13 - 4\dfrac{1}{3} =$ _____

13. $5\dfrac{1}{4} - 1\dfrac{1}{2} =$ _____

Problem Solving

Use the table for Exercises 14 and 15. The table shows the average lengths of some insects in the United States.

Insect	Length (in.)
Monarch Butterfly	$3\frac{1}{2}$
Walking Stick	4
Grasshopper	$1\frac{3}{4}$
Bumble Bee	$\frac{5}{8}$

14. **Mathematical PRACTICE 2 Reason** Is the difference in length between a Monarch butterfly and a bumble bee greater or less than the difference in length between a walking stick and grasshopper? Explain your reasoning.

_____

15. Find the difference in length between a walking stick and a bumble bee.

_____

16. Arlo had 5 gallons of paint. He used $2\frac{5}{8}$ gallons. How much paint does he have left?

_____

My Work!

HOT Problems

17. **Mathematical PRACTICE 1 Plan Your Solution** Write a subtraction problem in which you have to rename a fraction and whose solution is between 3 and 4.

_____

18. **? Building on the Essential Question** How are equivalent fractions used in renaming when subtracting?

_____

_____

MY Homework

Homework Helper

Need help? connectED.mcgraw-hill.com

Find $2 - 1\frac{1}{4}$.

Estimate $2 - 1 = 1$

You cannot subtract $\frac{1}{4}$ from 0 fourths.

Rename 2 as $1\frac{4}{4}$ to show more fourths.

$$2 \rightarrow 1\frac{4}{4}$$ Rename 2 as $1\frac{4}{4}$.

$$\begin{array}{r} 2 \rightarrow 1\frac{4}{4} \\ -\ 1\frac{1}{4} \rightarrow -\ 1\frac{1}{4} \\ \hline \frac{3}{4} \end{array}$$

Subtract the wholes.
$1 - 1 = 0$
Subtract the fractions.
$\frac{4}{4} - \frac{1}{4} = \frac{3}{4}$

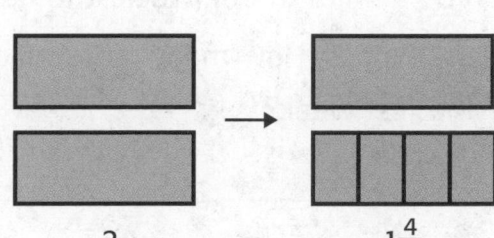

$$2 \quad = \quad 1\frac{4}{4}$$

So, $2 - 1\frac{1}{4} = \frac{3}{4}$.

Check for Reasonableness $1 \approx \frac{3}{4}$

Practice

Estimate, then subtract. Write each difference in simplest form.

1. $\begin{array}{r} 2\frac{1}{8} \\ -\ 1\frac{7}{8} \\ \hline \end{array}$

2. $\begin{array}{r} 12\frac{1}{4} \\ -\ 5\frac{2}{3} \\ \hline \end{array}$

3. $8\frac{1}{6} - 3\frac{5}{6} = $ _____

Problem Solving

Mathematical
4. PRACTICE ➊ **Make a Plan** Sherman's backpack weighs $6\frac{1}{4}$ pounds. Brie's backpack weighs $5\frac{3}{4}$ pounds. How much heavier is Sherman's backpack than Brie's backpack?

My Work!

Mathematical
5. PRACTICE ➋ **Use Number Sense** Veronica jogged $10\frac{3}{16}$ miles in one week. The next week she jogged $8\frac{7}{16}$ miles. How many more miles did she jog the first week?

Mathematical
6. PRACTICE ➏ **Be Precise** Careta swam $7\frac{1}{8}$ miles. Joey swam $5\frac{5}{8}$ miles. How many more miles did Careta swim than Joey?

Come in Joey!
The water is fine.

Test Practice

7. Ross has 6 yards of material. He bought $2\frac{1}{3}$ more yards. Then he used $6\frac{5}{6}$ yards. How many yards of material does he have left?

Ⓐ $1\frac{1}{2}$ yards

Ⓒ $3\frac{2}{3}$ yards

Ⓑ $3\frac{1}{6}$ yards

Ⓓ $8\frac{1}{3}$ yards

Vocabulary Check

Write each of the following on the lines to make a true sentence.

equivalent fractions   least common denominator   like fractions

mixed number   unlike fractions

1. Fractions that have the same denominator, such as $\frac{1}{3}$ and $\frac{2}{3}$, are

_____.

2. Fractions that have the same value, such as $\frac{1}{4}$ and $\frac{2}{8}$, are

_____.

3. The _____ of $\frac{5}{6}$ and $\frac{7}{12}$ is 12.

4. A number that has a whole number part and a fraction part is

called a _____

5. An example of two _____ are $\frac{1}{8}$ and $\frac{3}{4}$.

Concept Check

Round each fraction to 0, $\frac{1}{2}$, or 1.

6. $\frac{4}{7} \approx$ _____

7. $\frac{9}{10} \approx$ _____

8. $\frac{2}{9} \approx$ _____

Add. Write each sum in simplest form.

9. $\frac{5}{9} + \frac{1}{9} = $ _____

10. $\frac{6}{7} + \frac{1}{7} = $ _____

11. $\frac{5}{8} + \frac{1}{8} = $ _____

12. $\frac{3}{5} + \frac{1}{10} = $ _____

13. $\frac{1}{2} + \frac{1}{8} = $ _____

14. $\frac{2}{7} + \frac{5}{14} = $ _____

15. $6\frac{3}{4} + 2\frac{2}{4} = $ _____

16. $1\frac{1}{12} + 3\frac{2}{12} = $ _____

17. $12\frac{1}{3} + 6\frac{2}{6} = $ _____

Estimate, then subtract. Write each difference in simplest form.

18. $\frac{7}{16} - \frac{3}{16} = $ _____

19. $\frac{11}{12} - \frac{7}{12} = $ _____

20. $\frac{4}{5} - \frac{2}{5} = $ _____

21. $\frac{8}{9} - \frac{5}{6} = $ _____

22. $\frac{11}{12} - \frac{7}{8} = $ _____

23. $\frac{7}{10} - \frac{1}{5} = $ _____

Problem Solving

24. Russ is putting his vacation photographs in an album that is $12\frac{1}{8}$ inches long and $10\frac{1}{4}$ inches wide. Should he trim the edges of the photographs to 12 inches long and 10 inches wide or to $12\frac{1}{2}$ inches long and $10\frac{1}{2}$ inches wide?

25. Steve watched television for $\frac{3}{4}$ hour on Monday and $\frac{5}{6}$ hour on Tuesday. How much longer did he watch television on Tuesday than on Monday?

26. When Ricki walks to school on the sidewalk, she walks $\frac{7}{10}$ mile. She then takes a shortcut across the field, which is $\frac{1}{4}$ mile long. How long is Ricki's route to school?

Test Practice

27. Peta was swimming with stingrays. The first stingray she swam with was $5\frac{1}{4}$ feet long. The second one she swam with was $4\frac{3}{4}$ feet long. How much longer was the first stingray?

    Ⓐ $\frac{1}{2}$ foot   Ⓒ $1\frac{1}{4}$ feet

    Ⓑ $\frac{3}{4}$ foot   Ⓓ $1\frac{1}{2}$ feet

Reflect

Use what you learned about fraction operations to complete the graphic organizer.

Write an Example

Real-World Example

ESSENTIAL QUESTION

How can equivalent fractions help me add and subtract fractions?

Estimate

Vocabulary

Now reflect on the ESSENTIAL QUESTION Write your answer below.

_____

_____

Chapter 10 Multiply and Divide Fractions

In My Kitchen

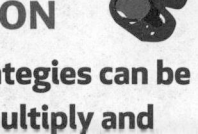
ESSENTIAL QUESTION

What strategies can be used to multiply and divide fractions?

Watch

Watch a video!

699

MY Standards

Number and Operations – Fractions

5.NF.4 Apply and extend previous understandings of multiplication to multiply a fraction or whole number by a fraction.

5.NF.4a Interpret the product $\left(\frac{a}{b}\right) \times q$ as a parts of a partition of q into b equal parts; equivalently, as the result of a sequence of operations $a \times q \div b$.

5.NF.4b Find the area of a rectangle with fractional side lengths by tiling it with unit squares of the appropriate unit fraction side lengths, and show that the area is the same as would be found by multiplying the side lengths. Multiply fractional side lengths to find areas of rectangles, and represent fraction products as rectangular areas.

5.NF.5 Interpret multiplication as scaling (resizing), by:

5.NF.5a Comparing the size of a product to the size of one factor on the basis of the size of the other factor, without performing the indicated multiplication.

5.NF.5b Explaining why multiplying a given number by a fraction greater than 1 results in a product greater than the given number (recognizing multiplication by whole numbers greater than 1 as a familiar case); explaining why multiplying a given number by a fraction less than 1 results in a product smaller than the given number; and relating the principle of fraction equivalence $\frac{a}{b} = \frac{(n \times a)}{(n \times b)}$ to the effect of multiplying $\frac{a}{b}$ by 1.

5.NF.6 Solve real world problems involving multiplication of fractions and mixed numbers, e.g., by using visual fraction models or equations to represent the problem.

5.NF.7 Apply and extend previous understandings of division to divide unit fractions by whole numbers and whole numbers by unit fractions.

5.NF.7a Interpret division of a unit fraction by a non-zero whole number, and compute such quotients.

5.NF.7b Interpret division of a whole number by a unit fraction, and compute such quotients.

5.NF.7c Solve real world problems involving division of unit fractions by non-zero whole numbers and division of whole numbers by unit fractions, e.g., by using visual fraction models and equations to represent the problem.

Standards for Mathematical PRACTICE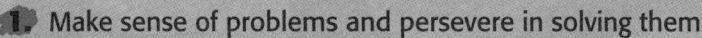

1. Make sense of problems and persevere in solving them.
2. Reason abstractly and quantitatively.
3. Construct viable arguments and critique the reasoning of others.
4. Model with mathematics.
5. Use appropriate tools strategically.
6. Attend to precision.
7. Look for and make use of structure.
8. Look for and express regularity in repeated reasoning.

= focused on in this chapter

Name _____

Am I Ready?

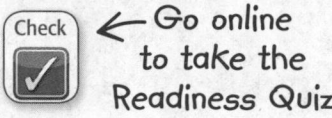 Check ✓ ← Go online to take the Readiness Quiz

Estimate. Use rounding. Show how you estimated.

1. $8\frac{5}{6} + 7\frac{1}{5} \approx$ $\frac{89}{11}$ $\frac{53}{6} + \frac{36}{5} = \frac{89}{11}$ 2. $17\frac{1}{4} - 3\frac{7}{8} \approx$ $\frac{45}{4}$ $\frac{69}{4} + \frac{24}{8} = \frac{4}{4}$

3. Tyra filled the two water coolers shown with flavored sports drink. About how many gallons of sports drink does she have altogether?

$\frac{32}{12}$

$\frac{12}{8} + \frac{20}{4} = \frac{32}{12}$

$4\frac{7}{8}$ gal $5\frac{1}{4}$ gal

Estimate. Use rounding or compatible numbers. Show how you estimated.

4. $196 \times 12 \approx$ 2352

5. $79 \times 31 \approx$ 2449

6. $304 \div 6 \approx$?

7. $558 \div 8 \approx$?

Add or subtract. Write in simplest form.

8. $\frac{1}{6} + \frac{1}{6} =$ $\frac{2}{12} = \left(\frac{1}{6}\right)$

9. $\frac{3}{4} + \frac{2}{5} =$ $\left(\frac{5}{9}\right) - \left(\frac{5}{9}\right)$

10. $\frac{8}{9} - \frac{5}{9} =$ $\frac{3}{0} = \left(\frac{3}{1}\right)$

11. $\frac{7}{12} - \frac{1}{12} =$ $\frac{6}{0} = \left(\frac{6}{1}\right)$

Shade the boxes to show the problems you answered correctly.

How Did I Do? → | 1 | 2 | 3 | 4 | 5 | 6 | 7 | 8 | 9 | 10 | 11 |

MY Math Words
Vocab $^{a}b_{c}$

Review Vocabulary

decimal point   denominator   digit
divide   equivalent   greatest common factor (GCF)
least common multiple (LCM)   mixed numbers   multiply
number line

Making Connections
Use the Venn diagram to categorize the review vocabulary.

Fractions   **Decimals**

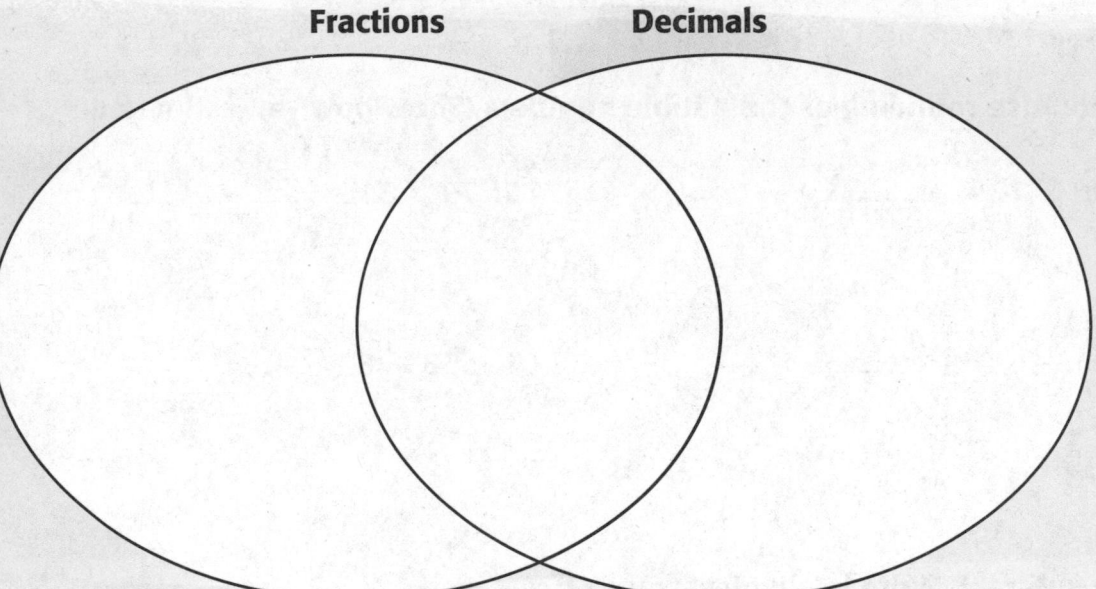

How is adding and subtracting fractions similar to adding and
subtracting whole numbers?

_____

_____

_____

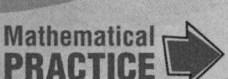
✂

Lesson 10-8

scaling

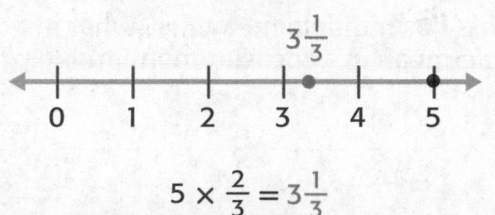

$$5 \times \frac{2}{3} = 3\frac{1}{3}$$

Lesson 10-9

unit fraction

$$\frac{1}{3} \leftarrow \boxed{\text{numerator} = 1}$$

Ideas for Use

- During the year, create a separate stack of cards for key math verbs, such as *scaling*. They will help you in problem solving.
- Use the blank cards to write review vocabulary cards.

- Group like ideas you find throughout the chapter, such as the relationship between multiplication and division. Use the blank cards to write notes about these concepts.

A fraction with a numerator of 1.

The Latin root *fract* means "break." How does this relate to the meaning of *fraction*?

_____

_____

_____

The process of resizing a number when it is multiplied by a fraction that is greater than or less than 1.

Scale has multiple meanings. What might *scale* mean in a measurement problem?

_____

_____

MY Foldable

FOLDABLES® Follow the steps on the back to make your Foldable.

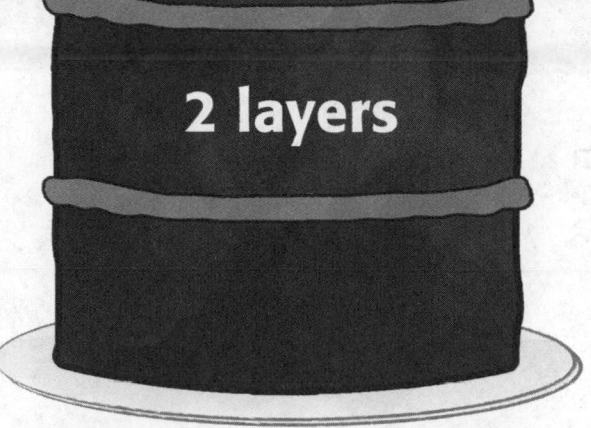

1 layer

$\frac{3}{4}$ cup sugar

$\frac{1}{3}$ cup butter

1 egg

$\frac{1}{2}$ teaspoon vanilla

$1\frac{1}{3}$ cups flour

$1\frac{1}{4}$ teaspoons baking powder

$\frac{1}{2}$ teaspoon salt

$\frac{3}{4}$ cup milk

2 layers

_____ cups sugar

_____ cup butter

_____ eggs

_____ teaspoon vanilla

_____ cups flour

_____ teaspoons baking powder

_____ teaspoon salt

_____ cups milk

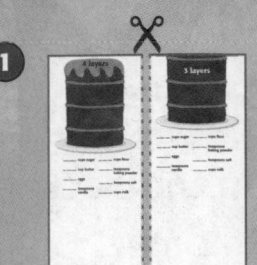
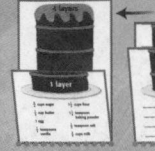

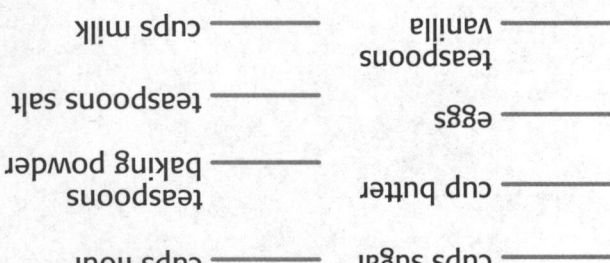

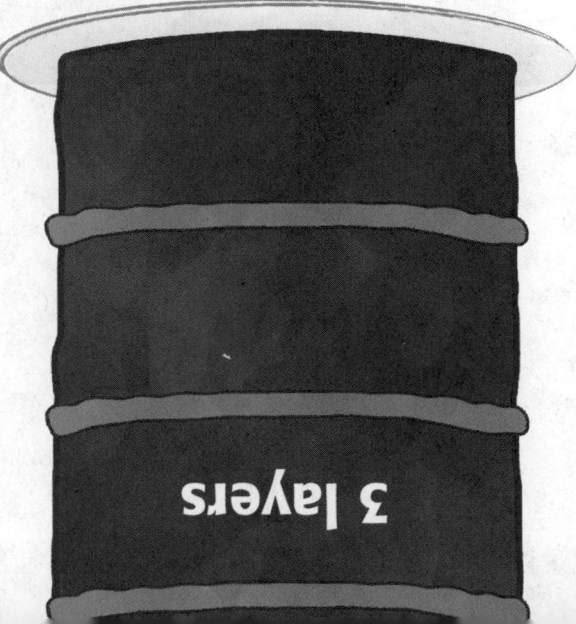

cups flour —

cups sugar —

teaspoons baking powder —

cup butter —

teaspoons salt —

eggs —

cups milk —

teaspoons vanilla —

3 layers

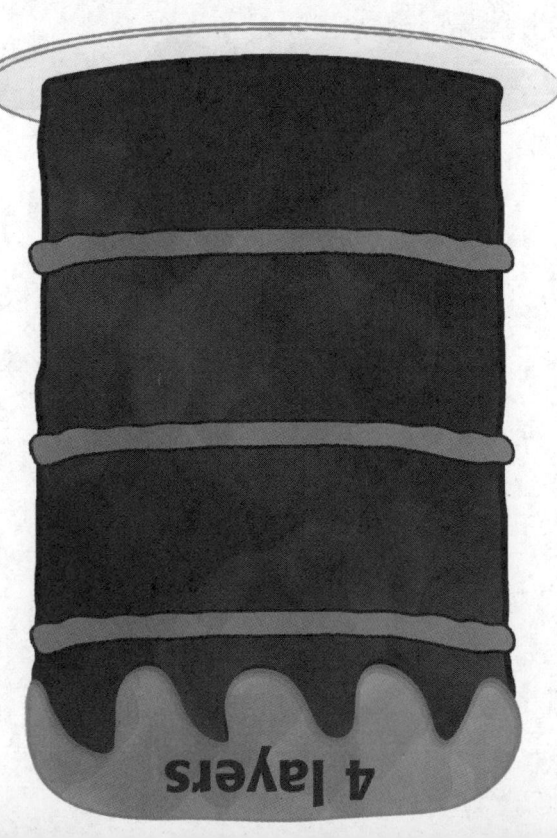

cups flour —

cups sugar —

teaspoons baking powder —

cups butter —

teaspoons salt —

eggs —

cups milk —

teaspoons vanilla —

4 layers

Name _____

Hands On
Part of a Number

Lesson 1

ESSENTIAL QUESTION
What strategies can be used to multiply and divide fractions?

You can use bar diagrams to find parts of a number.

$\frac{1}{4}$ of 16 is the same as $16 \div 4$

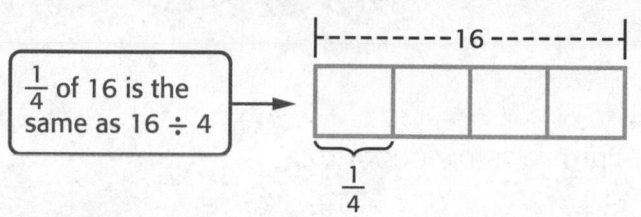

├────── 16 ──────┤

$\frac{1}{4}$

Take me out to the ball game!

Mathlete!

Draw It

Lelah threw 16 pitches in the first inning of a softball game. Of the pitches she threw, $\frac{3}{4}$ of them were strikes. How many strikes did she throw in the first inning?

Find $\frac{3}{4}$ of 16.

 The bar diagram represents the number of pitches she threw.

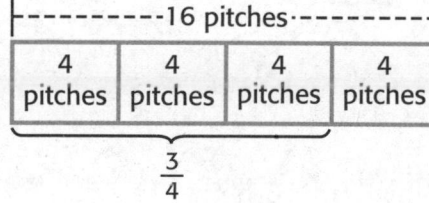

├────── 16 pitches ──────┤

| 4 pitches | 4 pitches | 4 pitches | 4 pitches |

$\frac{3}{4}$

2 Since the denominator is 4, the bar diagram was divided into

_____ equal sections.

Each section of the bar represents _____ pitches.

3 Use the diagram to determine $\frac{3}{4}$ of 16.

$4 + 4 + 4 =$ _____

$\frac{3}{4}$ of 16 is _____ .

So, Lelah threw _____ strikes.

Try It

Find $\frac{1}{3}$ of 15 using a bar diagram.

 Label the bar diagram that represents 15.

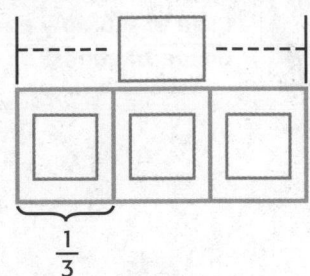

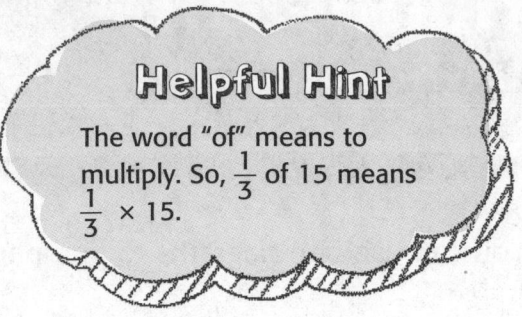

Helpful Hint

The word "of" means to multiply. So, $\frac{1}{3}$ of 15 means $\frac{1}{3} \times 15$.

 Since the denominator is 3, the bar diagram was divided into

_____ equal sections.

Each section of the bar represents _____.

 Use the diagram to determine $\frac{1}{3}$ of 15.

$\frac{1}{3}$ of 15 is the same as $15 \div 3$, which is the same as $\frac{1}{3} \times 15$.

What is $\frac{1}{3}$ of 15? _____

So, $\frac{1}{3}$ of 15 = _____.

Talk About It

1. **Mathematical PRACTICE 6 Explain to a Friend** Explain why $\frac{1}{4}$ of 16 is the same as $16 \div 4$.

_____

_____

_____

2. Explain why $\frac{3}{4}$ of 16 is the same as $3 \times 16 \div 4$.

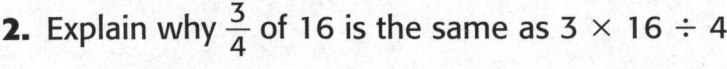

_____

Practice It

Draw a bar diagram to find each product.

3. $12 \times \dfrac{1}{2} =$ _____

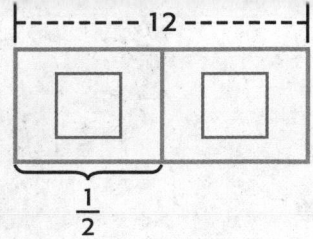

4. $\dfrac{2}{3}$ of $15 =$ _____

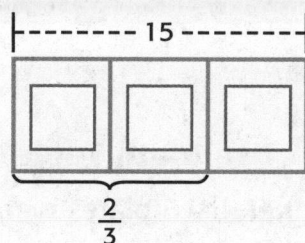

5. $\dfrac{2}{3}$ of $18 =$ _____

6. $9 \times \dfrac{1}{3} =$ _____

7. $8 \times \dfrac{1}{4} =$ _____

8. $\dfrac{1}{2}$ of $16 =$ _____

9. $25 \times \dfrac{2}{5} =$ _____

10. $24 \times \dfrac{3}{4} =$ _____

Apply It

Draw a bar diagram to help solve Exercises 11 and 12.

11. Leon used plant fertilizer on $\frac{4}{7}$ of his potted flowers. If he has 28 potted flowers, on how many did he use plant fertilizer?

_____

12. **Mathematical PRACTICE** 5 **Use Math Tools** Jeremy washed $\frac{3}{8}$ of the plates from dinner. If 16 plates were used, how many plates did Jeremy wash?

_____

My Drawing!

13. **Mathematical PRACTICE** 4 **Model Math** Write a real-world problem that could represent the bar diagram shown.

_____

_____

_____

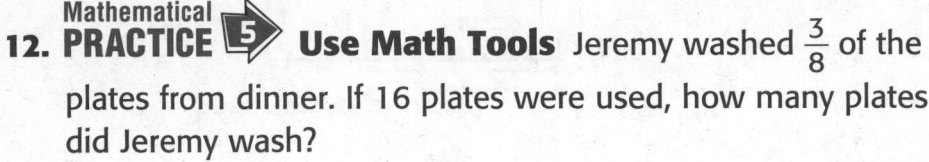

Write About It

14. How can I use models to find part of a number?

_____

_____

MY Homework

Lesson 1

Hands On: Part of a Number

Homework Helper

Need help? connectED.mcgraw-hill.com

Find $\frac{3}{4}$ of 40 using a bar diagram.

1 The bar diagram represents 40.

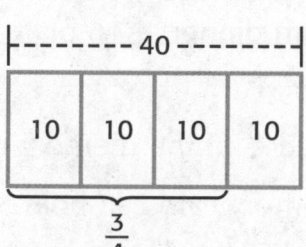

2 Since the denominator is 4, the bar diagram is divided into 4 equal sections.

Each section of the bar represents 10.

3 Use the diagram to determine $\frac{3}{4}$ of 40.

Since $\frac{1}{4}$ of 40 is the same as 40 ÷ 4, or 10, then $\frac{3}{4}$ of 40 is the same as 3 × 40 ÷ 4, or 30.

So, $\frac{3}{4}$ of 40 = 30.

Practice

Draw a bar diagram to find each product.

1. $\frac{2}{3}$ of 38 = _____

My Drawing!

2. $35 \times \frac{3}{5}$ = _____

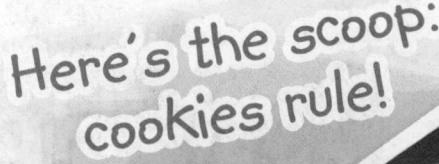

Problem Solving

Draw a bar diagram to help solve Exercises 3–5.

3. Hope used $\frac{1}{3}$ of the flour in the container to make cookies. If the container holds 12 cups of flour, how many cups did Hope use?

Here's the scoop: cookies rule!

_____

4. Elijah used $\frac{3}{4}$ of the memory on his cell phone memory card. If the memory card can hold 32 gigabytes, how many gigabytes did Elijah use?

_____

5. **Mathematical** **PRACTICE** **4** **Model Math** Jeremy used $\frac{5}{6}$ of a loaf of bread throughout the week. If there were 24 slices of bread, how many slices did Jeremy use?

_____

6. Write a real-world problem that could represent the bar diagram shown.

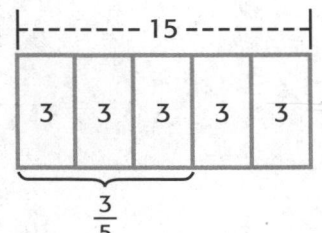

_____

_____

_____

Estimate Products of Fractions

Lesson 2

ESSENTIAL QUESTION
What strategies can be used to multiply and divide fractions?

You have already estimated products of decimals using rounding and compatible numbers. You can use the same strategies to estimate the products of fractions.

SNACK TO GO

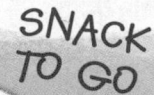

 ## Math in My World

Example 1

About $\frac{1}{3}$ of the containers of yogurt in a box are strawberry. If the box contains 17 containers, about how many containers are strawberry?

Estimate $\frac{1}{3} \times 17$.

Use compatible numbers.

$\frac{1}{3} \times 17 \approx \frac{1}{3} \times$ _____

> What number is close to 17 and also compatible with 3?

Complete the bar diagram to find the product of $\frac{1}{3} \times$ _____.

How many yogurt containers are represented in each section of the bar diagram? _____

So, $\frac{1}{3} \times 18 =$ _____.

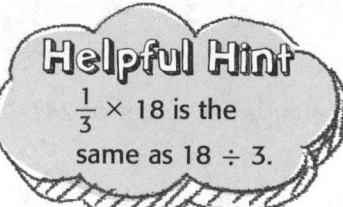

Helpful Hint
$\frac{1}{3} \times 18$ is the same as $18 \div 3$.

About _____ yogurt containers are strawberry.

Online Content at connectED.mcgraw-hill.com

Example 2

Estimate $\frac{2}{5} \times \frac{6}{7}$.

Use a number line to round each fraction to 0, $\frac{1}{2}$, or 1.

$\frac{2}{5} \times \frac{6}{7} \approx \boxed{} \times 1$

$\approx \boxed{}$ **Identity Property**

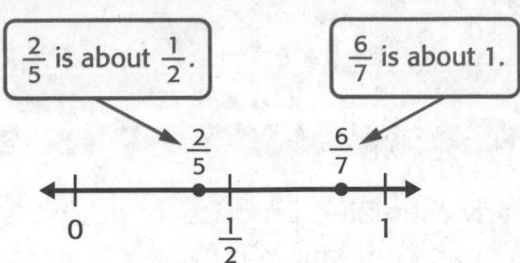

$\frac{2}{5}$ is about $\frac{1}{2}$.

$\frac{6}{7}$ is about 1.

So, $\frac{2}{5} \times \frac{6}{7}$ is about $\boxed{}$.

Example 3

Estimate $7\frac{2}{7} \times 3\frac{7}{8}$.

Round each mixed number to the nearest whole number.

Round $7\frac{2}{7}$ down to _____ and round $3\frac{7}{8}$ up to _____.

$7\frac{2}{7} \times 3\frac{7}{8} \approx$ _____ × _____

$\approx$ _____

So, $7\frac{2}{7} \times 3\frac{7}{8}$ is about _____.

Talk MATH

Explain how you would estimate the product of $\frac{4}{5} \times \frac{5}{6}$.

Guided Practice

1. Estimate the product of $\frac{1}{2} \times 25$.

   Use compatible numbers.

   $\frac{1}{2} \times 25 \approx \frac{1}{2} \times$ _____   Think: What is half of 24?

   $\approx$ _____

   So, $\frac{1}{2} \times 25$ is about _____.

714 Chapter 10 Multiply and Divide Fractions

Independent Practice

Estimate each product. Draw a bar diagram if necessary.

2. $\frac{2}{3} \times 13$ _____

3. $\frac{1}{3} \times 20$ _____

4. $\frac{1}{2} \times 33$ _____

5. $17 \times \frac{1}{4}$ _____

6. $\frac{7}{8} \times \frac{1}{9}$ _____

7. $\frac{3}{5} \times \frac{8}{9}$ _____

8. $\frac{1}{6} \times \frac{5}{7}$ _____

9. $\frac{1}{4} \times \frac{8}{9}$ _____

10. $2\frac{2}{3} \times 3\frac{1}{6}$ _____

11. $6\frac{4}{5} \times 5\frac{7}{8}$ _____

12. $10\frac{1}{7} \times 4\frac{4}{5}$ _____

13. $2\frac{6}{7} \times 6\frac{2}{9}$ _____

Problem Solving

14. A cup of chocolate chips weighs about 9 ounces. A recipe calls for $3\frac{3}{4}$ cups of chocolate chips. About how many ounces of chocolate chips are needed?

_____

15. **Mathematical** **PRACTICE** 5 **Use Math Tools** Isabella sent out 22 invitations to her birthday party. If about $\frac{1}{4}$ of the invitations are for her school friends, about how many invitations are for her school friends? Draw a bar diagram to help you solve.

_____

My Work!

HOT Problems

16. **Mathematical** **PRACTICE** 1 **Make a Plan** Write a real-world problem involving the multiplication of two mixed numbers whose product is about 14. Then solve the problem.

_____

_____

17. **Building on the Essential Question** Explain when estimation would not be the best method for solving a problem.

_____

_____

Hands On
Model Fraction Multiplication

You can multiply a whole number by a fraction using repeated addition.

Draw It

Find $\frac{1}{3} \times 3$. Use repeated addition.

1 Since the denominator is 3, each model is divided into _____ equal sections.

2 Shade $\frac{1}{3}$ of each model.

How many sections of each model are shaded? _____

Add. How many total sections of the models are shaded? _____

Shade this number of sections on the last model.

3 The last model shows the product of $\frac{1}{3} \times 3$.

$\frac{1}{3} \times 3 = \dfrac{\boxed{}}{\boxed{}}$ ⟵ number of shaded sections

⟵ number of sections

What is $\frac{3}{3}$ equal to? _____

So, $\frac{1}{3} \times 3 = 3 \div 3$, or _____ .

Try It

Find the unknown in $2 \times \frac{3}{4} = \blacksquare$. Use models. Write in simplest form.

1 Divide each model below into _____ equal sections since the denominator is 4.

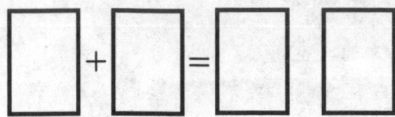

2 Shade $\frac{3}{4}$ of each of the first two models.

How many sections of each model are shaded? _____

Add. How many total sections of the models are shaded? _____

Shade this number of sections on the last two models.

3 The last two models show the product of $2 \times \frac{3}{4}$.

$2 \times \frac{3}{4} = \dfrac{\boxed{}}{\boxed{}}$ ←— number of shaded sections
←— number of sections

What is $\frac{6}{4}$ equal to?

So, $2 \times \frac{3}{4} = \boxed{}\,\dfrac{\boxed{}}{\boxed{}}$. ←— unknown

Helpful Hint
$2 \times \frac{3}{4}$ is the same as $3 \times 2 \div 4$.

Talk About It

1. Explain why $2 \times \frac{3}{4}$ is the same as $3 \times 2 \div 4$.

_____

_____

Mathematical
2. PRACTICE **2** **Reason** Explain how you could find $2 \times \frac{3}{4}$ without using models.

_____

_____

Practice It

Shade the models to find each product. Write in simplest form.

3. $4 \times \frac{1}{2} =$ _____

$\square + \square + \square + \square = \square\square$

4. $4 \times \frac{1}{3} =$ _____

$\square + \square + \square + \square = \square\square$

5. $\frac{2}{3} \times 5 =$ _____

$\square + \square + \square + \square + \square = \square\square\square\square$

6. $3 \times \frac{2}{3} =$ _____

$\square + \square + \square = \square\square$

Algebra Find each unknown. Shade the models to find each product. Write in simplest form.

7. $\frac{1}{5} \times 6 = \blacksquare$

$\blacksquare =$ _____

$\square + \square + \square + \square + \square + \square = \square\square$

8. $\frac{?}{5} \times 6 = \blacksquare$

$\blacksquare =$ _____

$\square + \square + \square + \square + \square + \square = \square\square\square$

Apply It

Algebra Use models to help you solve Exercises 9 and 10. Then complete the equation.

9. Brandon saved 2 gigabytes of music on his MP3 player. Of the music saved, $\frac{1}{4}$ is hip hop. What fraction of a gigabyte did Brandon use for his hip hop music?

_____

Equation: $2 \times \frac{1}{4} = 2 \div$ _____ , or $\dfrac{\boxed{}}{\boxed{}}$

10. **Mathematical PRACTICE** 4 **Model Math** Over the past 6 hours, Natalie spent $\frac{1}{5}$ of each hour kneading bread. How much time did she spend kneading bread in all?

_____

Equation: $6 \times \frac{1}{5} = 6 \div$ _____ , or $\boxed{}\dfrac{\boxed{}}{\boxed{}}$

Rise to the top!

11. **Mathematical PRACTICE** 1 **Plan Your Solution** Write and solve a real-world problem that could be represented by the model below.

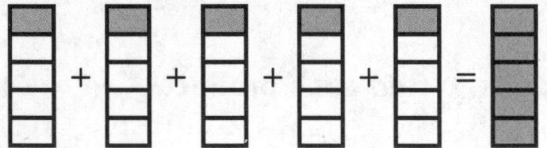

_____

_____

Write About It

12. Explain why $\frac{2}{3} \times 12$ can be written as $2 \times 12 \div 3$.

_____

_____

MY Homework

Homework Helper

Need help? connectED.mcgraw-hill.com

Find $5 \times \frac{2}{5}$ using models. Write in simplest form.

1 Each model below is divided into 5 equal sections since the denominator is 5.

☐ + ☐ + ☐ + ☐ + ☐ = ☐ ☐

2 Of each model, $\frac{2}{5}$ is shaded. There are 2 sections of each model that are shaded.
Add. A total of 10 sections of the models are shaded.

3 The last two models show the product of $5 \times \frac{2}{5}$.

$5 \times \frac{2}{5} = \frac{10}{5}$ ← number of shaded sections
← number of sections

$= 2$ Simplify.

So, $5 \times \frac{2}{5} = 2$.

Helpful Hint
$5 \times \frac{2}{5}$ is the same
as $2 \times 5 \div 5$.

Practice

Shade the models to find each product. Write in simplest form.

1. $4 \times \frac{2}{3} =$ _____

☐ + ☐ + ☐ + ☐ = ☐ ☐ ☐

2. $\frac{3}{4} \times 2 =$ _____

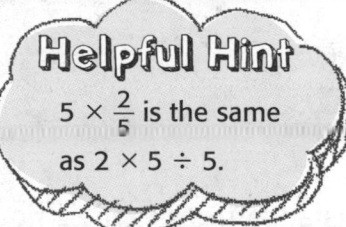

 **Problem Solving**

My Drawing!

3. Melinda spent 4 hours reviewing for her midterm exams. She spent $\frac{1}{4}$ of the time studying for social studies. How many hours did she spend on social studies?

_____

Equation: $4 \times \frac{1}{4} = 4 \div$ _____, or ☐

4. Cody wants to watch a new movie that is 3 hours long. He has watched $\frac{1}{6}$ of the movie so far. What fraction of one hour did Cody spend watching the new movie?

_____

Equation: $3 \times \frac{1}{6} = 3 \div$ _____, or ☐/☐

5. The distance from Paula's house to school is 5 miles. There is a set of railroad tracks one-fourth of the distance away from her house. How many miles away are the railroad tracks from Paula's house?

_____

Equation: $5 \times \frac{1}{4} = 5 \div$ _____, or ☐ ☐/☐

You're on track!

Multiply Whole Numbers and Fractions

Lesson 4

ESSENTIAL QUESTION
What strategies can be used to multiply and divide fractions?

You can write a whole number as a fraction.

$$12 \longrightarrow \frac{12}{1}$$

Math in My World

Example 1

Wild parrots spend $\frac{1}{6}$ of the day looking for food. How many hours a day does a parrot spend looking for food?

Find $\frac{1}{6} \times 24$. ◄── There are 24 hours in a day.

$\frac{1}{6} \times 24 = \frac{1}{6} \times \frac{24}{1}$ Write 24 as a fraction.

$= \dfrac{1 \times 24}{6 \times 1}$ ◄── Multiply the numerators.
◄── Multiply the denominators.

$= \dfrac{24}{6}$ or _____   Simplify.

So, a wild parrot spends _____ hours a day looking for food.

Check $\frac{1}{6} \times 24 = 1 \times 24 \div 6$, or _____

POLLY WANTS
A CRACKER!

Key Concept   Multiply Fractions

> To multiply a whole number by a fraction, write the whole number as a fraction. Then multiply the numerators and multiply the denominators.

Example 2

Find the unknown in $2 \times \frac{4}{5} = \blacksquare$.

Estimate $2 \times 1 =$ _____

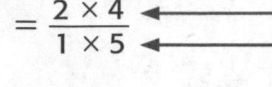

Helpful Hint

Multiplication is commutative.

$$2 \times \frac{4}{5} = \frac{4}{5} \times 2$$

$2 \times \frac{4}{5} = \frac{2}{1} \times \frac{4}{5}$ Write 2 as a fraction.

$= \dfrac{2 \times 4}{1 \times 5}$ ←— Multiply the numerators.
←— Multiply the denominators.

$= \dfrac{\square}{\square}$ or $\dfrac{\square \ \square}{\square}$ Simplify.

So, $2 \times \dfrac{4}{5} = \dfrac{\square}{\square}$ or $\dfrac{\square \ \square}{\square}$.   ←— (unknown)

Check $\dfrac{\square \ \square}{\square} \approx 2$

Talk MATH

Explain how you could find the product of 50 and $\frac{2}{5}$ mentally.

Guided Practice

1. Find $\dfrac{1}{2} \times 4$. Write in simplest form.

$\dfrac{1}{2} \times 4 = \dfrac{1}{2} \times \dfrac{4}{1}$ Write 4 as $\dfrac{4}{1}$.

$= \dfrac{1 \times 4}{2 \times 1}$ Multiply.

$= \dfrac{\square}{\square}$ or _____   Simplify.

So, $\dfrac{1}{2} \times 4 =$ _____ .

Independent Practice

Multiply. Write in simplest form.

2. $\frac{1}{3} \times 12 =$ _____

3. $\frac{1}{4} \times 20 =$ _____

4. $\frac{5}{6} \times 18 =$ _____

5. $\frac{1}{5} \times 7 =$ _____

6. $\frac{2}{3} \times 14 =$ _____

7. $\frac{2}{5} \times 11 =$ _____

8. $12 \times \frac{1}{6} =$ _____

9. $13 \times \frac{2}{13} =$ _____

10. $24 \times \frac{3}{4} =$ _____

Mathematical PRACTICE 2 **Use Algebra** **Find the unknown in each equation. Write in simplest form.**

11. $5 \times \frac{1}{4} = \blacksquare$

12. $15 \times \frac{7}{10} = \blacksquare$

13. $32 \times \frac{5}{6} = \blacksquare$

$\blacksquare =$ _____

$\blacksquare =$ _____

$\blacksquare =$ _____

Problem Solving

Nacho Cheese Dip
$\frac{1}{2}$ cup salsa

1 pound cheddar cheese

$\frac{1}{2}$ pound hamburger

14. Arleta is making nacho cheese dip for a party. She needs to make 5 batches. How much salsa will she need?

_____

15. Maria ate $\frac{1}{4}$ of a pizza. If there were 20 slices of pizza, how many slices did Maria eat?

_____

16. Mathematical **PRACTICE** 4 **Model Math** Write and solve a real-world problem that involves multiplying $8 \times \frac{5}{8}$.

_____

_____

HOT Problems

17. Mathematical **PRACTICE** 3 **Which One Doesn't Belong?** Circle the expression that does not belong with the other three. Explain.

| $\frac{1}{2} \times 12$ | $9 \times \frac{2}{3}$ | $\frac{1}{4} \times 20$ | $\frac{1}{6} \times 36$ |

_____

18. **Building on the Essential Question** Can all whole numbers be written as fractions? Explain.

_____

_____

Name ....

Hands On
Use Models to Multiply Fractions

Lesson 5

ESSENTIAL QUESTION
What strategies can be used to multiply and divide fractions?

Draw It

First $\frac{1}{3} \times \frac{1}{4}$. Write in simplest form.

To find $\frac{1}{3} \times \frac{1}{4}$, find the area of a $\frac{1}{3}$- by $\frac{1}{4}$-unit rectangle.

1 Divide the square into _____ equal rows since the denominator of the first fraction is 3.

2 Divide the square into _____ equal columns since the denominator of the second fraction is 4.

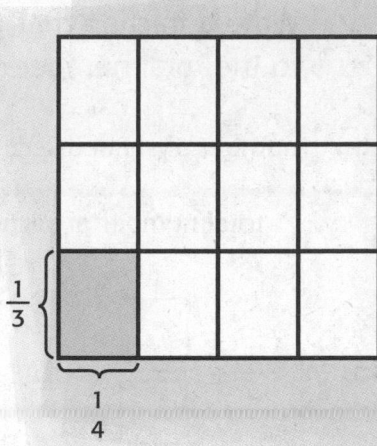

3 Shade the portion of the model where $\frac{1}{3}$ and $\frac{1}{4}$ intersect.

How many sections of the model are shaded? _____

4 Write a fraction that compares the number of shaded sections to the total number of sections.

$$\boxed{} \longleftarrow \text{number of shaded sections}$$

$$\boxed{} \longleftarrow \text{total number of sections}$$

So, $\frac{1}{3} \times \frac{1}{4} = \dfrac{\boxed{}}{\boxed{}}$.

Try It

Find $\frac{1}{2} \times \frac{2}{3}$. Write in simplest form.

To find $\frac{1}{2} \times \frac{2}{3}$, find the area of a $\frac{1}{2}$- by $\frac{2}{3}$-unit rectangle.

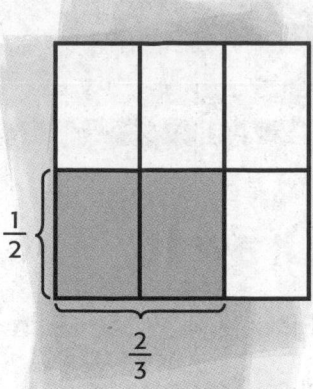

1 Divide the square into _____ equal rows since the denominator of the first fraction is 2.

2 Divide the square into _____ equal columns since the denominator of the second fraction is 3.

3 Shade the portion of the model where $\frac{1}{2}$ and $\frac{2}{3}$ intersect.

How many sections of the model are shaded? _____

4 Write a fraction that compares the number of shaded sections to the total number of sections. Simplify the fraction.

number of shaded sections ⟶ $\dfrac{\square}{\square} = \dfrac{\square}{\square}$ } Simplify.

total number of sections ⟶

So, $\frac{1}{2} \times \frac{2}{3} = \dfrac{\square}{\square}$.

Talk About It

1. In the second activity, how do the numerators of the fractions $\frac{1}{2}$ and $\frac{2}{3}$ relate to the total number of shaded sections, 2?

_____

2. In the second activity, how do the denominators of the fractions relate to the total number of sections, 6?

_____

Mathematical
3. PRACTICE 2 **Stop and Reflect** Write a rule you can use to multiply fractions without using models.

_____

Name

Practice It

Mathematical PRACTICE ⑤ **Use Math Tools** Shade the models to find each product. Write in simplest form.

4. $\frac{1}{6} \times \frac{2}{3} =$ _____

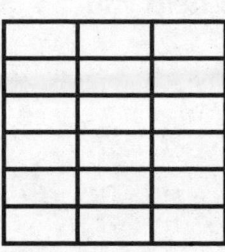

5. $\frac{3}{5} \times \frac{1}{3} =$ _____

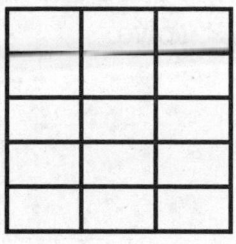

6. $\frac{3}{4} \times \frac{1}{6} =$ _____

7. $\frac{3}{5} \times \frac{3}{4} =$ _____

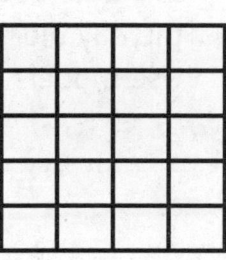

8. $\frac{3}{5} \times \frac{1}{5} =$ _____

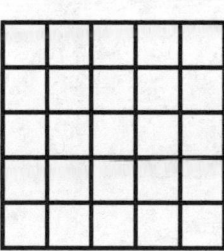

9. $\frac{1}{2} \times \frac{3}{5} =$ _____

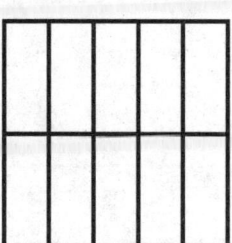

10. $\frac{3}{4} \times \frac{2}{5} =$ _____

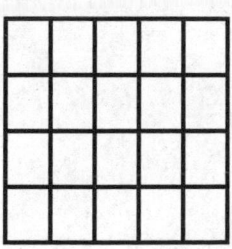

11. $\frac{2}{3} \times \frac{2}{3} =$ _____

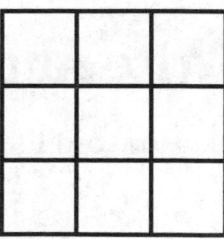

Apply It

12. Mathematical **PRACTICE** 5 **Use Math Tools** Ruby biked a trail that is $\frac{3}{5}$ mile each way. After biking $\frac{1}{4}$ of the trail, she stopped to rest. What fraction of a mile did Ruby bike before she stopped to rest? Use models to help you solve.

_____

My Work!

13. On Saturday, Tom spent $\frac{1}{3}$ of the day preparing snacks and decorating for a birthday party. Tom spent $\frac{3}{8}$ of this time decorating the cake. What fraction of the day did Tom spend decorating the cake? Use models to help you solve.

_____

14. Mathematical **PRACTICE** 4 **Model Math** Write a real-world problem that could be represented by the model. Then solve.

_____

_____

_____

_____

Write About It

15. How can I use models to multiply fractions?

_____

_____

Name _____

MY Homework

Homework Helper

Need help? connectED.mcgraw-hill.com

Find $\frac{2}{5} \times \frac{1}{2}$. Write in simplest form.

To find $\frac{2}{5} \times \frac{1}{2}$, find the area of a $\frac{2}{5}$- by $\frac{1}{2}$-unit rectangle.

1 The square is divided into 5 equal rows since the denominator of the first fraction is 5.

2 The square is divided into 2 equal columns since the denominator of the second fraction is 2.

3 The portion of the model where $\frac{2}{5}$ and $\frac{1}{2}$ intersect is shaded.

There are 2 sections of the model that are shaded.

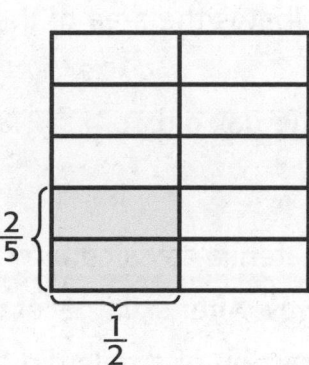

4 Write a fraction that compares the number of shaded sections to the total number of sections.

number of shaded sections $\longrightarrow$
total number of sections $\longrightarrow$ $\left. \dfrac{2}{10} = \dfrac{1}{5} \right\}$ Simplify.

So, $\frac{2}{5} \times \frac{1}{2} = \frac{2}{10}$ or $\frac{1}{5}$.

Practice

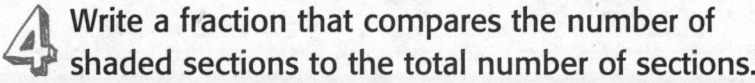
Shade the models to find each product. Write in simplest form.

1. $\frac{1}{2} \times \frac{1}{2} =$ _____

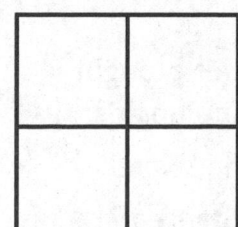

2. $\frac{4}{5} \times \frac{1}{2} =$ _____

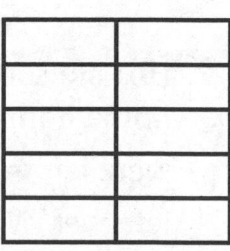

Problem Solving

Draw models to help you solve Exercises 3–7.

3. Charlotte spent $\frac{1}{2}$ of the day shopping at the mall. She spent $\frac{1}{4}$ of this time trying on jeans. What fraction of the day did Charlotte spend trying on jeans?

_____

4. Ms. Hendricks is buying a rectangular piece of land that is $\frac{2}{5}$ mile on one side and $\frac{5}{6}$ mile on an adjacent side. What is the area of the piece of land?

_____

5. Katenka cross country skied a trail that was $\frac{3}{4}$ mile each way. After skiing $\frac{2}{3}$ of the trail, she turned around. What fraction of a mile did Katenka ski before she turned around?

_____

6. **Mathematical**
   PRACTICE 5 **Use Math Tools** Merriam spent $\frac{3}{4}$ of her allowance at the mall. Of the money spent at the mall, $\frac{1}{3}$ was spent on new earrings. What part of her allowance did Merriam spend on earrings?

_____

7. Lexi ate some of the apples that her mother brought home from the farmer's fair. One-half of the apples were left over. Starr ate $\frac{1}{6}$ of the apples that were left over. What fraction of the apples did Starr eat?

_____

Name

Multiply Fractions

Lesson 6

ESSENTIAL QUESTION
What strategies can be used to multiply and divide fractions?

Multiplying fractions is similar to multiplying a whole number and a fraction. To multiply fractions, multiply the numerators and multiply the denominators.

$$\frac{3}{5} \times \frac{1}{2} = \frac{3 \times 1}{5 \times 2}$$

Get it while it's hot!

Math in My World

Watch Tutor

Example 1

Franco ate some pizza. Two-thirds of a pizza is left. Diendra ate $\frac{3}{4}$ of the leftover pizza. How much of a whole pizza did Diendra eat?

Find $\frac{3}{4}$ of $\frac{2}{3}$. ← $\frac{3}{4}$ of $\frac{2}{3}$ is the same as $\frac{3}{4} \times \frac{2}{3}$.

One Way Multiply first. Then simplify.

$$\frac{3}{4} \times \frac{2}{3} = \frac{3 \times 2}{4 \times 3}$$

$$= \frac{6}{12} \qquad \text{Multiply.}$$

$$= \frac{6 \div 6}{12 \div 6} = \frac{\boxed{}}{\boxed{}} \qquad \text{Simplify.}$$

Another Way Simplify first.

$$\frac{3}{4} \times \frac{2}{3} = \frac{\overset{1}{\cancel{3}} \times \overset{1}{\cancel{2}}}{\underset{2}{\cancel{4}} \times \underset{1}{\cancel{3}}}$$

Simplify.
Divide 3 and 3 by 3.
Divide 4 and 2 by 2.

$$= \frac{\boxed{}}{\boxed{}} \qquad \text{Multiply.}$$

So, Diendra ate $\dfrac{\boxed{}}{\boxed{}}$ of the whole pizza.

To find the area of a rectangle, multiply the length and the width.

Example 2

Find the area of the rectangle with a length of $\frac{3}{4}$ inch and a width of $\frac{5}{9}$ inch.

One Way Tile the rectangle as part of a unit square.

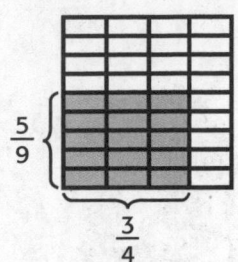

Each tile is $\frac{1}{4}$ in. by $\frac{1}{9}$ in., or $\frac{1}{36}$ in². The area of the unit square is 1 in².

Count the shaded tiles.
There are 15 tiles shaded.

$$15 \times \frac{1}{36} = \frac{15}{36}$$

$$= \frac{15 \div 3}{36 \div 3} = \frac{\square}{\square}$$ Simplify.

Another Way Multiply the side lengths.

$$\frac{3}{4} \times \frac{5}{9} = \frac{3 \times 5}{4 \times 9}$$

$$= \frac{15}{36}$$ Multiply.

$$= \frac{15 \div 3}{36 \div 3} = \frac{\square}{\square}$$ Simplify.

So, the area of the rectangle is $\frac{\square}{\square}$ square inch.

Talk MATH

Will the product of $\frac{2}{9} \times \frac{1}{3}$ be the same as the product of $\frac{2}{9} \times \frac{2}{6}$? Explain.

Guided Practice

1. Find $\frac{1}{7} \times \frac{1}{2}$. Write in simplest form.

$$\frac{1}{7} \times \frac{1}{2} = \frac{1 \times 1}{7 \times 2}$$

$$= \frac{\square}{\square}$$

So, $\frac{1}{7} \times \frac{1}{2} = \frac{\square}{\square}$.

MY Homework

Lesson 6
Multiply Fractions

Homework Helper

Need help? connectED.mcgraw-hill.com

Two-thirds of the vegetables that Michael planted were green.
Two-fifths of the green vegetables were peppers. What fraction
of the vegetable garden is green peppers? Write in simplest form.

Find $\frac{2}{3} \times \frac{2}{5}$.

$$\frac{2}{3} \times \frac{2}{5} = \frac{2 \times 2}{3 \times 5}$$

$$= \frac{4}{15} \qquad \text{Multiply.}$$

So, $\frac{4}{15}$ of the garden is green peppers.

Practice

Multiply. Write in simplest form.

1. $\frac{4}{9} \times \frac{3}{8} =$ _____

2. $\frac{3}{4} \times \frac{5}{8} =$ _____

3. $\frac{1}{8} \times \frac{3}{4} -$ _____

4. $\frac{3}{5} \times \frac{1}{3} =$ _____

5. $\frac{5}{7} \times \frac{1}{4} =$ _____

6. $\frac{7}{8} \times \frac{2}{9} =$ _____

Problem Solving

ORANGE you glad I didn't say BANANA?

Algebra Write an equation to solve Exercises 7 and 8.

7. Carrie ate some of the oranges that her mother brought home from the grocery store. Of the oranges her mother bought, $\frac{1}{4}$ were left over. Shyla, Carrie's sister, ate $\frac{2}{5}$ of the oranges that were left over. What fraction of the oranges did Shyla eat?

_____

8. Find the area of the rectangle shown.

_____

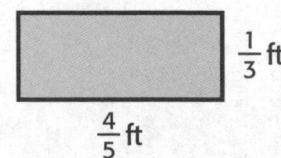

$\frac{1}{3}$ ft

$\frac{4}{5}$ ft

Mathematical
9. PRACTICE **7** **Identify Structure** Complete the steps to show how $\frac{4}{5} \times \frac{5}{6}$ can also be found using properties and equivalent fractions.

$\frac{4}{5} \times \frac{5}{6} = \frac{4 \times 5}{5 \times 6}$ Multiply numerators.
     Multiply denominators.

$= \frac{5 \times 4}{5 \times 6}$ _____ Property

$= \left[\frac{5}{5}\right] \frac{\times 4}{\times 6}$ $\frac{5}{5} = 1$

$= \frac{4}{6}$ _____ Property

$= \frac{2}{3}$ Simplify.

Test Practice

10. James made flash cards for $\frac{5}{6}$ of the vocabulary words. Lauretta made $\frac{1}{2}$ of the amount of cards that James made. What fraction of the vocabulary words did Lauretta make cards for?

   Ⓐ $\frac{5}{12}$ Ⓒ $\frac{7}{12}$

   Ⓑ $\frac{1}{2}$ Ⓓ $\frac{3}{4}$

Multiply Mixed Numbers

Lesson 7

ESSENTIAL QUESTION
What strategies can be used to multiply and divide fractions?

Math in My World

Example 1

A blueberry muffin recipe calls for $\frac{1}{2}$ cup of blueberries. A blueberry pie recipe calls for $3\frac{1}{2}$ times more blueberries. How many cups of blueberries are needed to make the pie?

1 The model shows $\frac{1}{2} \times 3\frac{1}{2}$. Shade the squares that represent the product.

How many squares did you shade? _____

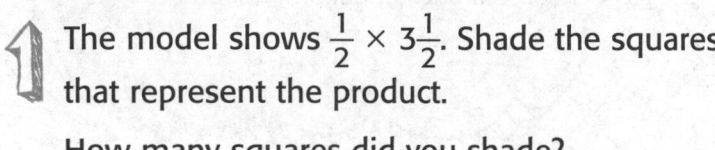

2 The shaded squares can be rearranged to form part of the model to the right.

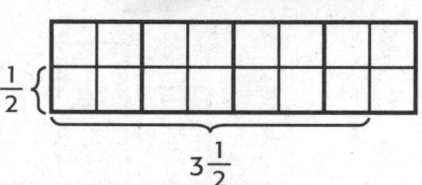

3 Write the product as an improper fraction.

$\boxed{}$ ← number of shaded sections

$\boxed{}$ ← number of squares in each section of the model

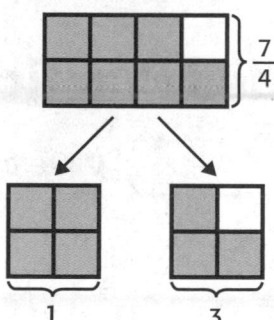

As a mixed number, this is $1\frac{3}{4}$.

So, the blueberry pie recipe calls for $\boxed{}\,\frac{\boxed{}}{\boxed{}}$ cups of blueberries.

Key Concept Multiply Mixed Numbers

> To multiply mixed numbers, write the mixed numbers as improper fractions. Then multiply as with fractions.

Example 2

Find the unknown in $1\frac{1}{2} \times 3\frac{3}{4} = $ ■.

$$1\frac{1}{2} \times 3\frac{3}{4} = \frac{\square}{\square} \times \frac{\square}{\square}$$ Write each mixed number as an improper fraction.

$$= \frac{\square}{\square}$$ Multiply.

$$= \frac{\square}{\square}$$ Simplify.

So, $1\frac{1}{2} \times 3\frac{3}{4} = \frac{\square}{\square}$.

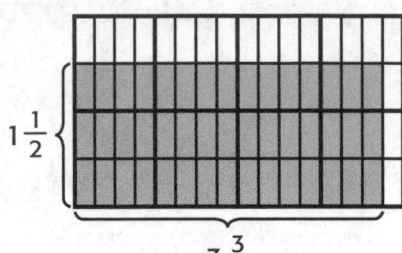

$1\frac{1}{2}$

$3\frac{3}{4}$

Helpful Hint

To write a mixed number as an improper fraction, multiply the denominator by the whole number and add the numerator. Keep the original denominator.

$$1\frac{1}{2} \rightarrow 2 \times 1 + 1 = \frac{3}{2}$$

Guided Practice ✓Check

1. Find $4\frac{1}{5} \times \frac{1}{2}$. Write in simplest form.

$$4\frac{1}{5} \times \frac{1}{2} = \frac{21}{5} \times \frac{1}{2}$$

$$= \frac{21}{10}$$

$$= \square\frac{\square}{\square}$$

So, $4\frac{1}{5} \times \frac{1}{2} = \square\frac{\square}{\square}$.

Talk MATH

Explain how to find the product of two mixed numbers.

Independent Practice

Multiply. Write in simplest form.

2. $1\frac{1}{3} \times \frac{2}{3} =$ _____

3. $\frac{2}{5} \times 2\frac{3}{4} =$ _____

4. $3\frac{3}{5} \times \frac{1}{4} =$ _____

5. $2\frac{1}{2} \times 4\frac{1}{5} =$ _____

6. $1\frac{1}{3} \times 3\frac{2}{3} =$ _____

7. $4\frac{2}{5} \times 1\frac{3}{4} =$ _____

8. $2\frac{4}{5} \times 6\frac{1}{8} =$ _____

9. $1\frac{5}{6} \times 4\frac{1}{3} =$ _____

10. $2\frac{3}{5} \times 3\frac{7}{8} =$ _____

Algebra Write a multiplication equation represented by each model.
Shade the product on the model.

11.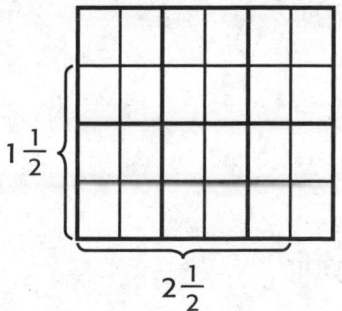

$1\frac{1}{2}$

$2\frac{1}{2}$

_____

12.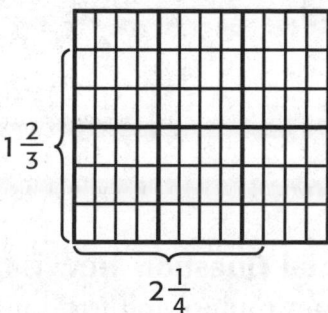

$1\frac{2}{3}$

$2\frac{1}{4}$

_____

13.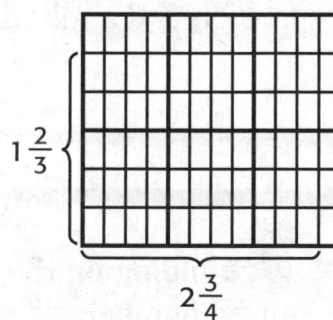

$1\frac{2}{3}$

$2\frac{3}{4}$

_____

Problem Solving

Mathematical
14. PRACTICE 4 **Model Math** Alexandra made a rectangular quilt that measured $3\frac{1}{4}$ feet in length by $2\frac{3}{4}$ feet in width. To find the area, multiply the length and width. What is the area of the quilt in square feet? Write an equation to solve.

_____

15. Laney bought $1\frac{2}{3}$ pounds of grapes. She also bought bananas that were $2\frac{1}{4}$ times the weight of the grapes. How much did the bananas weigh?

_____

16. Mrs. Plesich has a $16\frac{1}{2}$ ounce bag of chocolate chips. She will only use $\frac{1}{4}$ of the bag for decorating a cake. How many ounces will she use for decorating?

_____

My Work!

HOT Problems

Mathematical
17. PRACTICE 2 **Use Number Sense** Write and solve a real-world problem that involves finding the product of a fraction and a mixed number.

_____

_____

18. **Building on the Essential Question** How is multiplying mixed numbers different than multiplying fractions?

_____

_____

_____

MY Homework

Homework Helper

Need help? connectED.mcgraw-hill.com

A swimming pool has two diving boards. The shorter diving board is $2\frac{1}{4}$ yards high. The taller diving board is $1\frac{1}{3}$ the height of the shorter board. How tall is the taller diving board?

Find $2\frac{1}{4} \times 1\frac{1}{3}$.

$2\frac{1}{4} \times 1\frac{1}{3} = \frac{9}{4} \times \frac{4}{3}$ Write each mixed number as an improper fraction.

$= \frac{36}{12}$ Multiply.

$= 3$ Simplify.

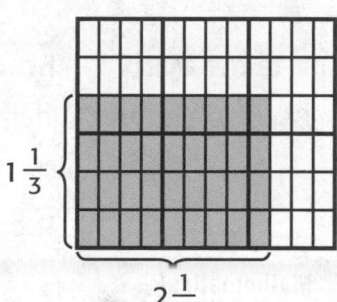

So, the taller diving board is 3 yards high.

Practice

Multiply. Write in simplest form.

1. $5\frac{1}{3} \times 1\frac{1}{4} = $ _____

2. $1\frac{2}{5} \times 3\frac{1}{6} = $ _____

3. $7\frac{1}{8} \times 2\frac{5}{6} = $ _____

Problem Solving

Oh so cheesy!

4. The table shows some ingredients in lasagna. If you make three times the recipe, how many cups of cheese are needed?

Tomato Sauce	Chopped Onion	Cheese
$3\frac{1}{2}$ cups	$\frac{1}{4}$ cup	$2\frac{2}{3}$ cups

_____

5. Karyn purchased a square picture frame. Each side measures $1\frac{1}{4}$ feet. What is the area of the picture frame in square feet?

_____

6. It takes Marty $1\frac{1}{4}$ hours to get ready for school. If $\frac{1}{5}$ of that time is used to shower, what fraction of an hour does it take him to shower?

_____

Mathematical
7. **PRACTICE 2 Use Algebra** Kallisto built a rectangular sign that measured $2\frac{3}{4}$ feet in length by $1\frac{1}{2}$ feet in width. To find the area, multiply the length and width. What is the area of the sign in square feet? Write an equation to solve.

_____

Test Practice

8. Antoinette bought $2\frac{2}{3}$ pounds of grapes. If she bought bananas that weighed $1\frac{1}{4}$ times as much as the grapes, how much did the bananas weigh?

Ⓐ $2\frac{1}{6}$ pounds   Ⓒ $3\frac{1}{4}$ pounds

Ⓑ $3\frac{1}{3}$ pounds   Ⓓ $3\frac{1}{2}$ pounds

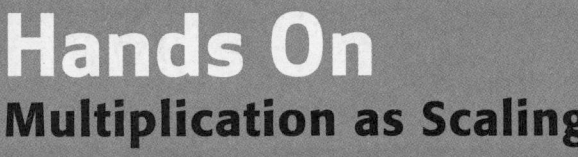

Hands On
Multiplication as Scaling

Lesson 8

ESSENTIAL QUESTION
What strategies can be used to multiply and divide fractions?

Scaling is the process of resizing a number when you multiply by a fraction that is greater than or less than 1.

Draw It

Multiply the number 2 by three fractions greater than 1.

1 Multiply the number 2 by three different fractions greater than 1, such as $1\frac{1}{5}$, $1\frac{1}{2}$, and $1\frac{3}{4}$.

$2 \times 1\frac{1}{5} = \boxed{} \frac{\boxed{}}{\boxed{}}$ $2 \times 1\frac{1}{2} = \boxed{}$ $2 \times 1\frac{3}{4} = \boxed{} \frac{\boxed{}}{\boxed{}}$

2 Plot 2 on the number line. Then plot the products on the number line.

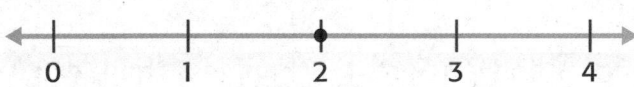

3 Compare the products. Circle whether the products are greater than, less than, or equal to 2.

greater than 2   less than 2   equal to 2

Multiplying a number by a fraction greater than one results

in a product that is _____ than the number.

Try It

Multiply the number 2 by three fractions less than 1.

 Multiply 2 by three different fractions less than 1 such as $\frac{1}{4}$, $\frac{1}{2}$, and $\frac{5}{8}$.

$$2 \times \frac{1}{4} = \frac{\square}{\square} \qquad 2 \times \frac{1}{2} = \square \qquad 2 \times \frac{5}{8} = \square \frac{\square}{\square}$$

Plot 2 on the number line. Then plot the products on the number line.

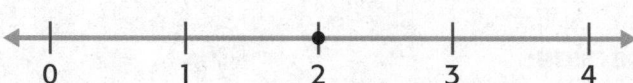

Compare the products. Circle whether the products are greater than, less than, or equal to 2.

greater than 2   less than 2   equal to 2

Multiplying a number by a fraction less than one results

in a product that is _____ than the number.

Talk About It

1. Predict whether the product of 3 and $\frac{4}{5}$ is greater than, less than, or equal to 3. Explain.

_____

2. **Mathematical PRACTICE** ③ **Draw a Conclusion** Predict whether the product of 2 and $2\frac{1}{5}$ is greater than, less than, or equal to 2. Explain.

_____

Practice It

Without multiplying, circle whether each product is greater than, less than, or equal to the whole number.

3. $2 \times \frac{1}{8}$

greater than

less than

equal to

4. $10 \times 1\frac{3}{5}$

greater than

less than

equal to

5. $1\frac{3}{4} \times 4$

greater than

less than

equal to

6. $12 \times \frac{5}{6}$

greater than

less than

equal to

7. $1\frac{1}{3} \times 2$

greater than

less than

equal to

8. $\frac{3}{5} \times 6$

greater than

less than

equal to

Algebra Without multiplying, circle whether the unknown in each equation is greater than, less than, or equal to the whole number.

9. $1\frac{2}{3} \times 4 = b$

greater than

less than

equal to

10. $8 \times \frac{4}{5} = h$

greater than

less than

equal to

11. $1 \times 5 = k$

greater than

less than

equal to

12. $6 \times \frac{1}{3} = n$

greater than

less than

equal to

 **Apply It**

For Exercises 13–15, analyze each product in the table.

First Factor	Second Factor	Product
$\frac{1}{2}$	$\frac{3}{4}$	$\frac{3}{8}$
1	$\frac{3}{4}$	$\frac{3}{4}$
$\frac{3}{2}$	$\frac{3}{4}$	$\frac{9}{8}$

13. Why is the first product less than $\frac{3}{4}$?

_____

14. Why is the product of 1 and $\frac{3}{4}$ equal to $\frac{3}{4}$?

_____

_____

15. Why is the product of $\frac{3}{2}$ and $\frac{3}{4}$ greater than $\frac{3}{4}$?

_____

Mathematical
16. PRACTICE 6 Explain to a Friend Miranda spent $\frac{1}{5}$ of her time cooking pasta. If she spent 2 hours cooking, did Miranda spend more than, less than, or equal to 2 hours cooking pasta? Explain.

_____

Mathematical
17. PRACTICE 3 Which One Doesn't Belong? Circle the multiplication expression that does not belong based on scaling. Explain.

| $\frac{2}{5} \times 3$ | $1\frac{1}{2} \times 3$ | $3 \times \frac{3}{5}$ | $3 \times \frac{1}{2}$ |

_____

_____

Write About It

18. How can I use scaling to help predict the product of a number and a fraction?

_____

_____

MY Homework

Homework Helper

Need help? connectED.mcgraw-hill.com

Multiply the number 3 by three fractions greater than 1 and three fractions less than 1.

1 Multiply 3 by three different fractions greater than 1 and three different fractions less than 1.

$$3 \times 1\frac{1}{5} = 3\frac{3}{5} \qquad 3 \times 1\frac{1}{2} = 4\frac{1}{2} \qquad 3 \times 1\frac{3}{4} = 5\frac{1}{4}$$

$$3 \times \frac{1}{4} = \frac{3}{4} \qquad 3 \times \frac{1}{2} = 1\frac{1}{2} \qquad 3 \times \frac{5}{8} = 1\frac{7}{8}$$

2 Plot 3 on the number line. Then plot the products.

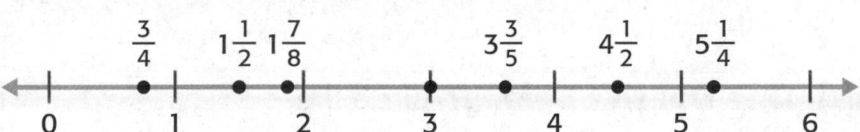

Multiplying a number by a fraction greater than one results in a product greater than the number.

Multiplying a number by a fraction less than one results in a product less than the number.

Practice

Without multiplying, circle whether each product is greater than, less than, or equal to the whole number.

1. $4 \times \frac{1}{7}$

greater than   less than   equal to

2. $12 \times 2\frac{5}{6}$

greater than   less than   equal to

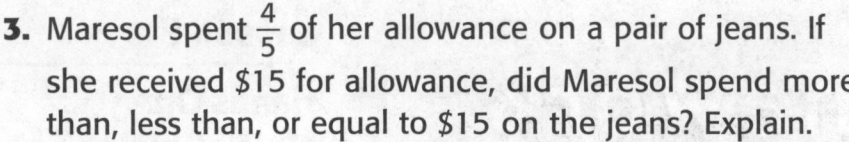

Problem Solving

Pocket money

3. Maresol spent $\frac{4}{5}$ of her allowance on a pair of jeans. If she received $15 for allowance, did Maresol spend more than, less than, or equal to $15 on the jeans? Explain.

_____

4. Dee used $2\frac{1}{3}$ cups of sugar for a cake recipe. If the amount of sugar the container holds is 3 times the amount she used, does the container hold more than, less than, or equal to 3 cups of sugar? Explain.

_____

5. Derek ran $\frac{3}{8}$ of a race before he had to stop for a break. If the race is 6 miles long, did Derek run more than, less than, or equal to 6 miles before he took a break? Explain.

_____

Mathematical
6. PRACTICE  **Which One Doesn't Belong?** Circle the multiplication expression that does not belong based on scaling. Explain.

$6 \times 2\frac{1}{2}$ $6 \times 1\frac{3}{5}$ $\frac{1}{2} \times 6$ $2\frac{2}{5} \times 6$

_____

_____

Vocabulary Check

7. Fill in the blank with the correct term or number to complete the sentence.

_____ is the process of resizing a number when you multiply by a fraction that is greater than or less than 1.

Hands On
Division with Unit Fractions

Lesson 9

ESSENTIAL QUESTION
What strategies can be used to multiply and divide fractions?

You know that 6 ÷ 3 means to take 6 objects and divide them into 3 equal groups. There would be 2 in each group. You can think of division with unit fractions in the same way. A **unit fraction** is a fraction with a numerator of 1.

Build It

Remy has 2 gallons of lemonoade. She needs to divide the lemonade into $\frac{1}{4}$-gallon containers. How many $\frac{1}{4}$-gallon containers does she need?

Find $2 \div \frac{1}{4}$. ⟵ THINK: How many groups of $\frac{1}{4}$ are in 2?

 Represent the 2 gallons by placing 2 whole fraction tiles.

 Place enough of the $\frac{1}{4}$-fraction tiles below the 2 whole fraction tiles to represent the same amount. Label the model with $\frac{1}{4}$-fraction tiles.

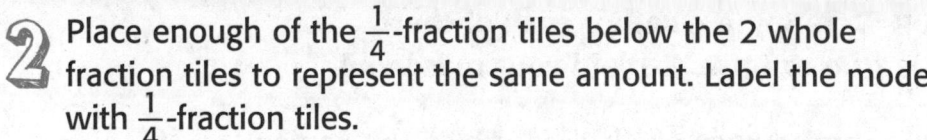

1				1

How many $\frac{1}{4}$-fraction tiles did it take to equal the 2 whole fractions tiles? _____

So, $2 \div \frac{1}{4} = $ _____.

Remy will need _____ containers.

Try It

Find $\frac{1}{5} \div 2$.

 Place a $\frac{1}{5}$-fraction tile.

Helpful Hint
$\frac{1}{5}$ represents one out of the five tiles that would make a whole.

2 Since you are dividing by 2, you are dividing $\frac{1}{5}$ into two equal groups.
Find a fraction tile that when placed twice, will be the same size as the $\frac{1}{5}$-fraction tile.
Label the model with the fraction tiles you placed.

What fraction tile did you place? _____

How many $\frac{1}{10}$-fraction tiles did it take to equal the $\frac{1}{5}$-fraction tile? _____

So, $\frac{1}{5} \div 2 = \dfrac{\boxed{}}{\boxed{}}$.

Talk About It

Mathematical PRACTICE 2 **Reason** Determine whether each of the following statements is *true* or *false*. Explain your reasoning.

1. When a whole number greater than one is divided by a unit fraction less than one, the quotient is always greater than the dividend.

_____

_____

2. When a unit fraction less than one is divided by a whole number greater than one, the quotient is always greater than the dividend.

_____

_____

Practice It

Mathematical PRACTICE 5 **Use Math Tools** Use fraction tiles to divide. Draw your models below.

3. $3 \div \frac{1}{3} =$ _____

4. $2 \div \frac{1}{5} =$ _____

5. $4 \div \frac{1}{2} =$ _____

6. $3 \div \frac{1}{6} =$ _____

7. $\frac{1}{3} \div 4 =$ _____

8. $\frac{1}{2} \div 3 =$ _____

9. $\frac{1}{3} \div 2 =$ _____

10. $\frac{1}{4} \div 2 =$ _____

11. **Mathematical** **PRACTICE** 5 **Use Math Tools** Randall is cutting a submarine sandwich into smaller pieces. He has 4 feet of sandwich that he wants to divide into $\frac{1}{3}$-foot sandwiches. How many sandwiches will he have after he cuts the sandwich? Use tiles to help you solve.

_____

12. Brittany is making party favors. She is dividing $\frac{1}{2}$ pound of jelly beans into 6 packages. How many pounds of jelly beans will be in each package? Use tiles to help you solve.

_____

13. **Mathematical** **PRACTICE** 2 **Use Symbols** Circle the missing divisor below from the equation $5 \div \blacksquare = 25$. Explain.

| $\frac{1}{2}$ | $\frac{2}{5}$ | 5 | $\frac{1}{5}$ |

_____

_____

14. **Building on the Essential Question** How can I use fraction tiles to help divide a whole number by a unit fraction?

_____

_____

MY Homework

Homework Helper

Need help? connectED.mcgraw-hill.com

Find $2 \div \frac{1}{8}$. ← THINK: How many groups of $\frac{1}{8}$ are in 2?

1 Represent the 2 by using 2 whole fraction tiles.

2 Place enough of the $\frac{1}{8}$-fraction tiles below the 2 whole fraction tiles to represent the same amount.

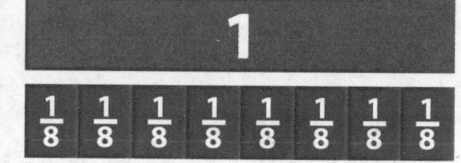

Sixteen $\frac{1}{8}$-fraction tiles were needed to equal the length of 2 whole fraction tiles.

So, $2 \div \frac{1}{8} = 16$.

Practice

Use fraction tiles to divide. Draw your models below.

1. $2 \div \frac{1}{6} =$ _____

2. $\frac{1}{4} \div 3 =$ _____

Problem Solving

3. Harold spent $\frac{3}{4}$ hour milking cows. He milked 3 groups of cows. If he spent an equal amount of time on each group of cows, what fraction of an hour did he spend milking each group? Use tiles to help you solve.

_____

4. Rosanne and her family drove for 8 hours on their vacation trip. Every $\frac{1}{2}$ hour, they played a different game. How many different games did they play? Use tiles to help you solve.

_____

5. **Mathematical** **PRACTICE** **5** **Use Math Tools** Girish used 4 cups of flour to make bread. He divided the flour into $\frac{1}{3}$-cup portions for each batch. How many batches of bread does Girish have? Use tiles to help you solve.

_____

6. Jason is cutting a roll of sausage into pieces that are $\frac{1}{2}$ inch thick. If the roll is 6 inches long, how many pieces of sausage can he cut? Use tiles to help you solve.

_____

7. Hubert cut a piece of yarn into pieces that are each $\frac{2}{3}$ foot long. If the yarn is 6 feet long, how many pieces of yarn did he cut? Use tiles to help you solve.

_____

My Drawing!

Hanging by a string!

Number and Operations – Fractions
5.NF.7, 5.NF.7b, 5.NF.7c

Divide Whole Numbers by Unit Fractions

Lesson 10

ESSENTIAL QUESTION
What strategies can be used to multiply and divide fractions?

You can divide a whole number by a unit fraction using models. A **unit fraction** is a fraction with a numerator of 1.

 ## Math in My World

Example 1

A sports Web site has score updates every $\frac{1}{4}$ hour. How many times does the Web site have score updates in a 3-hour period?

Find $3 \div \frac{1}{4}$ using models.   How many $\frac{1}{4}$s are in 3?

1 The model represents 3.

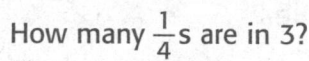

2 Divide each of the three rectangles into fourths.

3 Count the number of fourths.

There are _____ fourths in the model.

$3 \div \frac{1}{4} =$ _____ .

So, the Web site has _____ score updates in 3 hours.

Check You can check division problems using multiplication because they are inverse operations.

$12 \times \frac{1}{4} = \dfrac{\boxed{}}{\boxed{}}$ or _____

Example 2

Lisa takes four apple pies to her family reunion. Each pie is cut into sixths. How many pieces of pie can Lisa serve? Find the unknown in $4 \div \frac{1}{6} = g$.

 The model represents 4.

 Divide each of the four smaller rectangles into sixths.

 Count the number of sixths. There are _____ sixths in the model.

So, $4 \div \frac{1}{6} =$ _____ .

$g =$ _____ ◄——— unknown

Lisa can serve _____ pieces of pie.

Check Use multiplication to check your answer. _____ $\times \frac{1}{6} = 4$

Guided Practice

1. Find the quotient of $2 \div \frac{1}{3}$. Use a model. Check using multiplication.

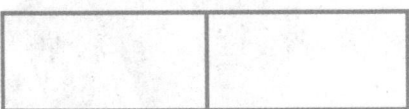

$2 \div \frac{1}{3} =$ _____

Check _____ $\times \frac{1}{3} = \dfrac{\square}{\square}$ or 2

Talk MATH

Why can you use multiplication to check your answer to a division problem?

Independent Practice

Find each quotient. Use a model. Check using multiplication.

2. $4 \div \frac{1}{3} =$ _____

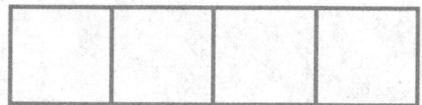

Check _____ $\times \frac{1}{3} = \dfrac{\boxed{}}{\boxed{}}$ or 4

3. $3 \div \frac{1}{5} =$ _____

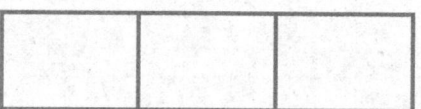

Check _____ $\times \frac{1}{5} = \dfrac{\boxed{}}{\boxed{}}$ or 3

4. $6 \div \frac{1}{4} =$ _____

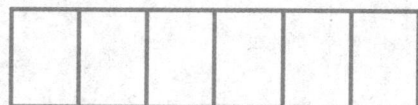

Check _____ $\times \frac{1}{4} = \dfrac{\boxed{}}{\boxed{}}$ or 6

5. $5 \div \frac{1}{4} =$ _____

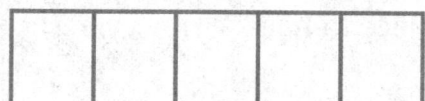

Check _____ $\times \frac{1}{4} = \dfrac{\boxed{}}{\boxed{}}$ or 5

6. $2 \div \frac{1}{2} =$ _____

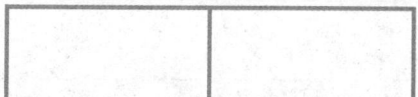

Check _____ $\times \frac{1}{2} = \dfrac{\boxed{}}{\boxed{}}$ or 2

7. $3 \div \frac{1}{6} =$ _____

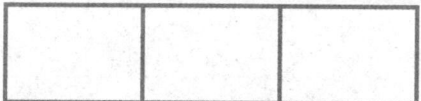

Check _____ $\times \frac{1}{6} = \dfrac{\boxed{}}{\boxed{}}$ or 3

Problem Solving

Mathematical
PRACTICE 5 **Use Math Tools** Draw a model to solve each problem.

8. Denise has 4 hours to paint crafts. She would like to spend no more than $\frac{1}{4}$ of each hour on each craft. How many crafts can she paint during that time?

_____

9. Laura has a 6-foot long piece of ribbon that she wants to cut to make bows. Each piece needs to be $\frac{1}{3}$-foot long to make one bow. How many bows will she be able to make?

_____

10. Ray has 8 pieces of wood to finish building a doghouse. How many pieces of wood will he have if he divides each piece in half?

_____

HOT Problems

Mathematical
11. **PRACTICE** 6 **Be Precise** Write and solve a real-world problem for $9 \div \frac{1}{8}$. Then explain the meaning of the quotient.

_____

_____

_____

12. **?** **Building on the Essential Question** Explain the relationship between division and multiplication.

_____

_____

MY Homework

Homework Helper

Need help? connectED.mcgraw-hill.com

The recipe Melinda is using to make fruit punch is for one serving. It calls for $\frac{1}{2}$ cup of pineapple juice. Melinda has 5 cups of pineapple juice. How many servings can Melinda make?

The model represents 5 cups of juice. Since each serving uses $\frac{1}{2}$ cup of pineapple juice, find how many $\frac{1}{2}$ cups are in 5 cups.

	5								
$\frac{1}{2}$	$\frac{1}{2}$	$\frac{1}{2}$	$\frac{1}{2}$	$\frac{1}{2}$	$\frac{1}{2}$	$\frac{1}{2}$	$\frac{1}{2}$	$\frac{1}{2}$	$\frac{1}{2}$

There are ten $\frac{1}{2}$-cup sections in the model, so $5 \div \frac{1}{2} = 10$.

Melinda can make 10 servings of fruit punch.

Check $10 \times \frac{1}{2} = \frac{10}{2}$ or 5

Practice

Find each quotient. Use a model. Check using multiplication.

1. $4 \div \frac{1}{5} =$ _____

Check _____ $\times \frac{1}{5} = \dfrac{\square}{\square}$ or 4

2. $6 \div \frac{1}{2} =$ _____

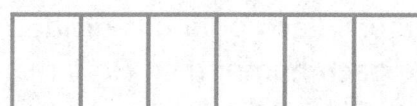

Check _____ $\times \frac{1}{2} = \dfrac{\square}{\square}$ or 6

Problem Solving

My Drawing!

3. Chef Erin has 4 pizzas that need to be cut into slices. Each slice represents $\frac{1}{6}$ of a pizza. How many slices of pizza can she serve?

_____

4. A baker cuts 3 cakes into pieces. Each piece represents $\frac{1}{8}$ of a cake. What is the total number of pieces?

_____

5. Mathematical PRACTICE 7 **Identify Structure** Circle the model below that represents $2 \div \frac{1}{3}$.

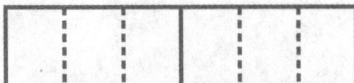 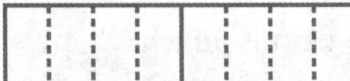

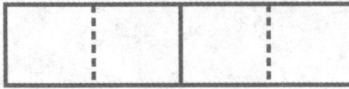 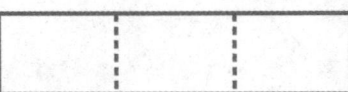

Test Practice

6. Karen uses five pounds of hamburger to grill hamburgers. Each pound is divided into thirds to make each hamburger. How many hamburgers can Karen serve? Find the unknown in $5 \div \frac{1}{3} = h$.

Ⓐ $h = 10$ hamburgers

Ⓑ $h = 12$ hamburgers

Ⓒ $h = 15$ hamburgers

Ⓓ $h = 18$ hamburgers

Divide Unit Fractions by Whole Numbers

Lesson 11

ESSENTIAL QUESTION
What strategies can be used to multiply and divide fractions?

 **Math in My World**  Tutor

Example 1

Elijah is sorting his music into playlists. One-half of his music is rock. He wants to make 3 separate rock playlists. If each playlist is the same size, what fraction of Elijah's music will be on one of the rock playlists?

Find $\frac{1}{2} \div 3$.

 The model shows the amount of Elijah's music that is rock.

rock

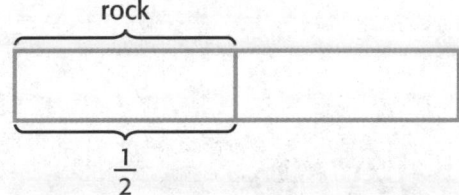

$\frac{1}{2}$

2 Divide each section into 3 equal parts.

How many equal sections are there now? _____

How many of the parts represent the fraction of the music that is on one of the rock playlists?

3 Write the fraction. ☐ ← one rock playlist section

☐ ← total sections

So, $\frac{1}{2} \div 3 = \dfrac{\square}{\square}$.

Check Use multiplication to check.

$$\frac{1}{6} \times 3 = \frac{1}{6} \times \frac{3}{1} = \frac{3}{6} \text{ or } \frac{1}{2}$$

Rock star!

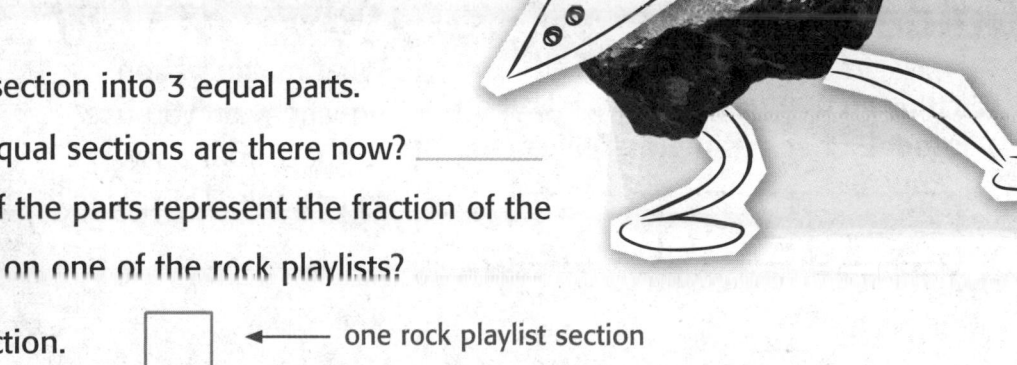

Example 2

In the buffet line, there was only $\frac{1}{4}$ of a pan of spaghetti left. Three friends decide to divide the spaghetti evenly. What fraction of a whole pan of spaghetti did each friend receive? Find the unknown in $\frac{1}{4} \div 3 = s$.

1 The model shows $\frac{1}{4}$.

$$\boxed{\quad}\boxed{\quad}\boxed{\quad}\boxed{\quad}$$
$$\underbrace{\qquad}_{\frac{1}{4}}$$

2 Divide each of the four sections into 3 equal parts. There are _____ total sections.

3 Write the fraction.

$\boxed{\quad}$ ⟵ one section

$\boxed{\quad}$ ⟵ total sections

So, $\frac{1}{4} \div 3 = \dfrac{\boxed{\quad}}{\boxed{\quad}}$.

$s = \dfrac{\boxed{\quad}}{\boxed{\quad}}$ ⟵ (unknown)

Each friend receives $\dfrac{\boxed{\quad}}{\boxed{\quad}}$ of a pan of spaghetti.

Guided Practice (Check ✓)

1. Find the quotient of $\frac{1}{3} \div 3$. Use the model. Check using multiplication.

$$\boxed{\quad}\boxed{\quad}\boxed{\quad}$$
$$\underbrace{\qquad}_{\frac{1}{3}}$$

$\frac{1}{3} \div 3 = \dfrac{\boxed{\quad}}{\boxed{\quad}}$

Check $\dfrac{\boxed{\quad}}{\boxed{\quad}} \times 3 = \dfrac{\boxed{\quad}}{\boxed{\quad}}$ or $\frac{1}{3}$

Talk MATH

What multiplication equation can you use to check your answer to Example 2? Explain.

Independent Practice

Find each quotient. Use each model. Check using multiplication.

2. $\frac{1}{2} \div 6 =$ _____

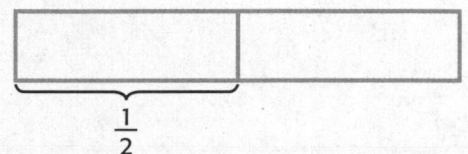

Check _____ $\times 6 =$ _____ or $\frac{1}{2}$

3. $\frac{1}{7} \div 2 =$ _____

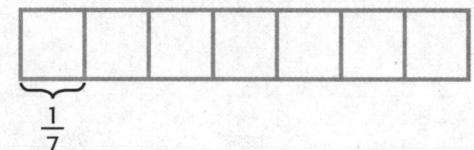

Check _____ $\times 2 =$ _____ or $\frac{1}{7}$

4. $\frac{1}{4} \div 4 =$ _____

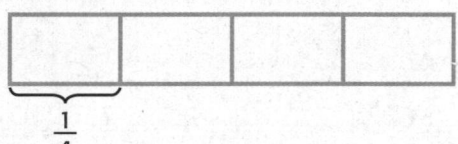

Check _____ $\times 4 =$ _____ or $\frac{1}{4}$

5. $\frac{1}{5} \div 2 =$ _____

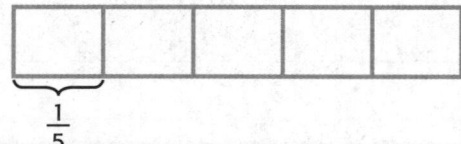

Check _____ $\times 2 =$ _____ or $\frac{1}{5}$

6. $\frac{1}{2} \div 4 =$ _____

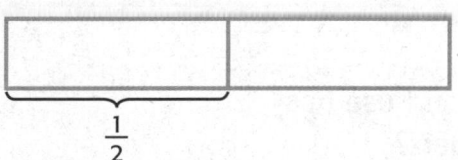

Check _____ $\times 4 =$ _____ or $\frac{1}{2}$

7. $\frac{1}{9} \div 2 =$ _____

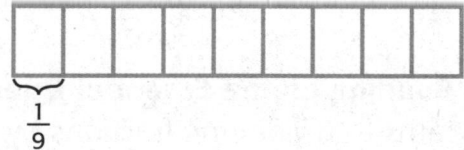

Check _____ $\times 2 =$ _____ or $\frac{1}{9}$

Problem Solving

Mathematical PRACTICE 5 **Use Math Tools** Draw a model to solve each problem.

8. One-fifth of the text messages that Natalie sends are to her family. She sends an equal amount of messages to her mom, dad, and sister. What fraction of all the text messages she sends go to her sister?

_____

9. Asher has $\frac{1}{2}$ ton of mulch to spread equally in 8 square yards. How many tons of mulch will be spread in each square yard?

_____

10. Hyun has $\frac{1}{3}$ of his science project to complete still. He wants to spend an equal amount of time on his science project over the next 3 nights. What fraction of the science project will he complete each night?

_____

HOT Problems

11. **Mathematical PRACTICE 6** **Be Precise** Write and solve a real-world problem for $\frac{1}{6} \div 4$. Then explain the meaning of the quotient.

_____

_____

_____

12. **Building on the Essential Question** How can I use bar diagrams to divide unit fractions by whole numbers?

_____

_____

_____

Name

MY Homework

Homework Helper [eHelp]

Need help? connectED.mcgraw-hill.com

Find $\frac{1}{6} \div 2$.

1 The model shows $\frac{1}{6}$.

2 Divide each of the six equal sections into 2 equal parts. There are now 12 sections.

3 Write the fraction.

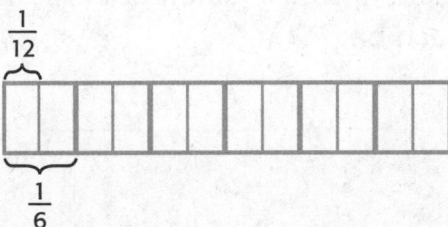

$\frac{1}{12}$ ← one section
  ← total sections

So, $\frac{1}{6} \div 2 = \frac{1}{12}$.

Check $\frac{1}{12} \times 2 = \frac{2}{12}$ or $\frac{1}{6}$

Practice

Find each quotient. Use each model. Check using multiplication.

1. $\frac{1}{2} \div 3 =$ _____

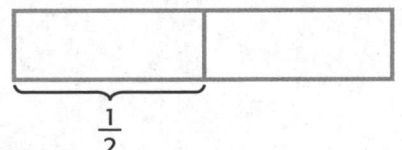

Check _____ × 3 = _____ or $\frac{1}{2}$

2. $\frac{1}{4} \div 5 =$ _____

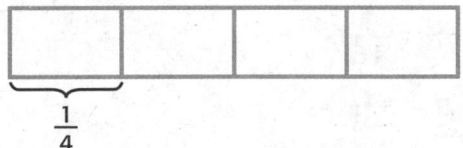

Check _____ × 5 = _____ or $\frac{1}{4}$

Problem Solving

Draw a model to solve Exercises 3 and 4.

3. There is only one granola bar left out of a pan of 10 bars. If Will and Rachel decide to split the last granola bar, what fraction of the entire pan of granola bars will each friend receive?

_____

4. **Mathematical PRACTICE 4** **Model Math** Tyson has $\frac{1}{2}$ pound of raisins to divide equally into 7 different bags. What fraction of a pound will be in each bag?

_____

5. **Mathematical PRACTICE 6** **Explain to a Friend** Write and solve a real-world problem for $\frac{1}{2} \div 3$. Then explain the meaning of the quotient.

_____

_____

_____

Test Practice

6. There is $\frac{1}{6}$ of a birthday cake left over. If 3 friends share the remaining cake equally, what fraction of the entire cake will each friend receive? Find the unknown in $\frac{1}{6} \div 3 = p$.

Ⓐ $p = \frac{1}{16}$

Ⓑ $p = \frac{1}{18}$

Ⓒ $p = \frac{1}{21}$

Ⓓ $p = \frac{1}{24}$

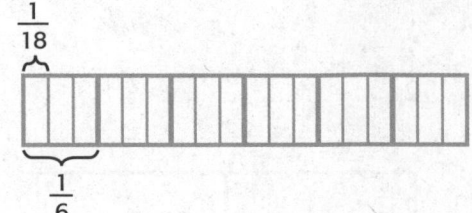

$\frac{1}{18}$

$\frac{1}{6}$

Number and Operations – Fractions
5.NF.4, 5.NF.4a, 5.NF.6, 5.NF.7, 5.NF.7b, 5.NF.7c

Problem-Solving Investigation
STRATEGY: Draw a Diagram

Lesson 12

ESSENTIAL QUESTION
What strategies can be used to multiply and divide fractions?

Learn the Strategy

Jaheim visited an aquarium over the weekend and saw 35 species of fish. This was $\frac{1}{5}$ the total number of species of fish. How many total species of fish are at the aquarium?

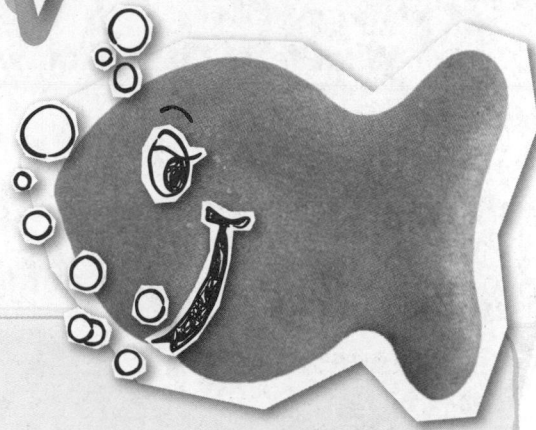

1 Understand

What facts do you know?

Jaheim saw _____ species of fish and this is _____ of the total amount of species.

What do you need to find?

• the total number of species of _____ in the aquarium

2 Plan

I can solve the problem by drawing a diagram.

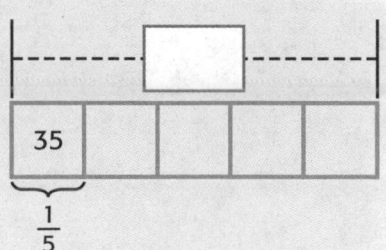

3 Solve

The model is divided into fifths. Since each part represents 35 species,

there is a total of 35 × 5, or _____ species of fish at the aquarium.

4 Check

Is my answer reasonable? _____ ÷ 5 = 35

Practice the Strategy

Victor has $18 in a piggy bank. He spends $\frac{2}{3}$ of the money on a video game and $\frac{1}{6}$ on candy. How much money will Victor have left?

Feed the PIG!

 Understand

What facts do you know?

_____

_____

What do you need to find?

_____

2 Plan

_____

3 Solve

 Check

Is my answer reasonable? Explain.

_____

Apply the Strategy

Solve each problem by drawing a diagram.

1. Mrs. Vallez purchased sand toys that were originally $20. She received $\frac{1}{4}$ off of the total price. How much did she save?

2. Sue has four DVDs and Terry has six DVDs. They put all their DVDs together and sold them for $10 for two DVDs. How much money will they earn if they sell all of their DVDs?

3. **Mathematical PRACTICE 4 Model Math** Jacinda is decorating cookies for a class party. She can decorate $\frac{2}{3}$ of a cookie per minute. At this rate, how many cookies can she decorate in 15 minutes?

4. At a bird sanctuary, Ricky counted 80 birds. Of the birds he counted, $\frac{1}{4}$ were baby birds. If he counted an equal number of adult males and females, how many adult female birds did Ricky count?

5. **Mathematical PRACTICE 5 Use Math Tools** To make chocolate ice cream, you need about $\frac{3}{8}$ pound of chocolate. How many pounds of chocolate will you need to make 4 batches of ice cream?

Use any strategy to solve each problem.
- Draw a diagram.
- Work backward.
- Guess, check, and revise.
- Act it out.

6. A cook needs 12 pounds of flour. He wants to spend the least amount of money. If a 2-pound bag costs $1.59 and a 5-pound bag costs $2.89, how many bags of each type of flour should he buy? What will be the total cost?

My Work!

Mathematical
7. PRACTICE 5 **Use Math Tools** A ride at a theme park lasts $1\frac{1}{2}$ minutes. It takes 2 minutes to prepare the ride for each trip. How many times can the ride be completed in 30 minutes?

8. On Monday, 21 DVDs were checked out at the library. This is 3 less than half the amount of books checked out that day. How many books were checked out?

Mathematical
9. PRACTICE 4 **Model Math** Leo takes 30 minutes to eat dinner, 15 minutes to change clothes, and 20 minutes to walk to practice. If Leo needs to be at hockey practice at 7:15 P.M., what time does he need to begin?

10. Mi-Ling sets up tables for her art class students. Each square table can seat two people on each side. How many people can be seated if 8 square tables are pushed together in a row?

MASTERPIECE!

MY Homework

Homework Helper  eHelp

Need help? connectED.mcgraw-hill.com

Norah earned $50 helping her neighbor. She spent $\frac{3}{5}$ of the money on a purse and put $\frac{1}{5}$ in the bank. How much money will Norah have left?

1 Understand

What facts do you know?

Norah earned $50 and spent $\frac{3}{5}$ on a purse and put $\frac{1}{5}$ in the bank.

What do you need to find?

the amount of money Norah has left

2 Plan

I can solve the problem by drawing a diagram.

3 Solve

The bar diagram represents the total amount earned.

There are 5 equal sections and each section represents $10.

So, Norah has $10 left.

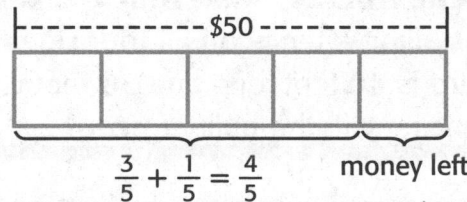

4 Check

Is my answer reasonable? Explain.

Use multiplication to check. $\frac{1}{5} \times 50 = 10$

Problem Solving

Solve each problem by drawing a diagram.

1. Paul is decorating cupcakes for a birthday party. He can decorate $\frac{3}{4}$ of a cupcake per minute. At this rate, how many cupcakes can he decorate in 12 minutes?

_____

Mathematical
2. **PRACTICE** 4 **Model Math** To make cheesecake, you need about $2\frac{1}{2}$ pounds of cream cheese. How many pounds of cream cheese will you need to make 2 cheesecakes?

_____

3. Scott viewed 72 classic cars at a car show. Of the cars he viewed, $\frac{3}{8}$ were sports cars. How many sports cars did Scott view?

_____

Mathematical
4. **PRACTICE** 5 **Use Math Tools** A builder is installing a fence on all four sides of a backyard. The yard is 40 feet long and 50 feet wide. How much fencing will the builder need?

_____

5. Misty purchased a pair of jeans that were originally $24, but were on sale for $\frac{1}{3}$ off of the total price. How much did she save?

_____

My Drawing!

Vocabulary Check

Read each clue. Fill the corresponding section of the crossword
puzzle to answer each clue. Use the words in the word bank.

dividend	divisor	fraction	inverse operations
quotient	mixed number	scaling	unit fraction

Across

1. The result of a division problem.

2. The number that divides the dividend.

3. A number that is being divided.

4. The process of resizing a number when
you multiply by a fraction that is greater
than or less than 1.

5. Operations that undo each other, such
as multiplication and division.

Down

6. A fraction with a numerator of 1.

7. A number that has a whole number
part and a fraction part.

8. A number that represents part
of a whole or part of a set.

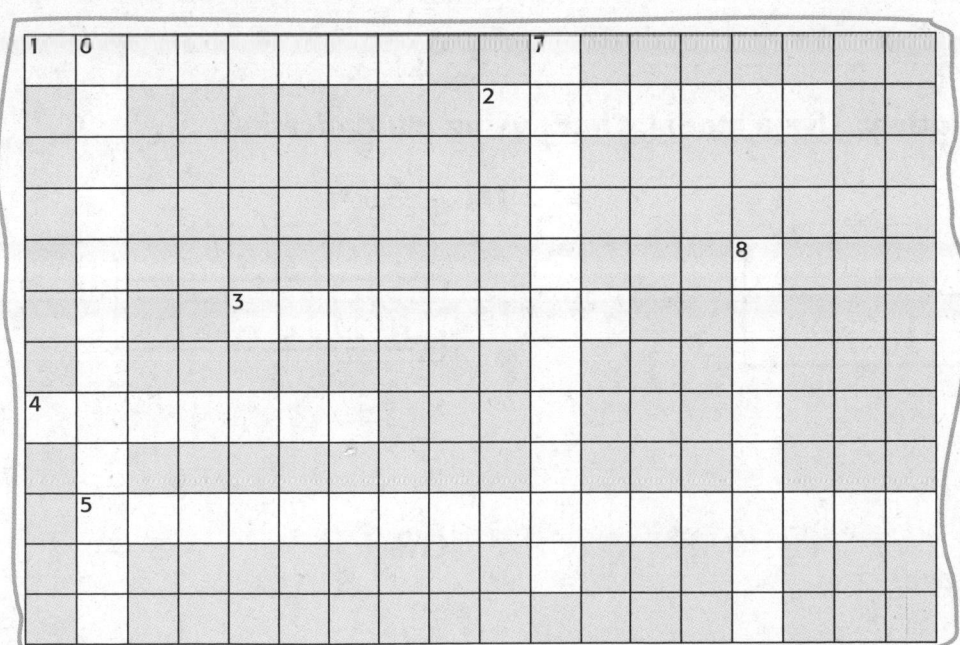

Concept Check ✓

Estimate each product. Draw a bar diagram if necessary.

9. $\frac{3}{4} \times 23$

_____

10. $\frac{1}{5} \times 22$

_____

Multiply. Write in simplest form.

11. $\frac{1}{3} \times 21 =$ _____

12. $26 \times \frac{1}{2} =$ _____

13. $\frac{1}{5} \times \frac{3}{8} =$ _____

14. $\frac{1}{2} \times \frac{7}{8} =$ _____

15. $3\frac{1}{4} \times \frac{3}{5} =$ _____

16. $1\frac{1}{8} \times 3\frac{2}{3} =$ _____

Find each quotient. Use a model. Check using multiplication.

17. $2 \div \frac{1}{6} =$ _____

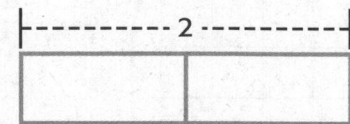

18. $\frac{1}{3} \div 5 =$ _____

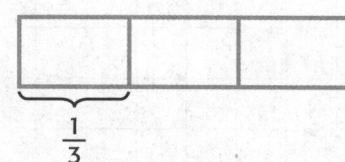

Check _____ $\times \frac{1}{6} =$ _____ or 2   **Check** _____ $\times 5 =$ _____ or $\frac{1}{3}$

Problem Solving

19. Ken is working on a project for social studies. He has a piece of poster board that needs to be divided equally into $\frac{1}{3}$-foot sections. The poster board is 4 feet wide. How many sections will Ken have on the poster board?

_____

20. Samantha measured the dimensions of a rectangular poster frame. It was $\frac{3}{4}$ yard long and $\frac{1}{2}$ yard wide. What is the area of the frame?

_____

21. At Middle Avenue School, $\frac{1}{30}$ of the students are on the track team. The track coach offered to buy pizza or subs for everyone on the team. Of the students for whom the coach bought food, $\frac{3}{5}$ ordered pizza. What fraction of the students at Middle Avenue School ate pizza?

_____

22. Saraid is dividing $\frac{1}{4}$ pound of cashews into 5 plastic bags. What fraction of a pound of cashews will be in each plastic bag? Draw a model to help you solve.

_____

Test Practice

23. The Great Rope Company sells a rope that is 5 feet long. Juan wants to make sections that are $\frac{1}{3}$ foot long. How many sections will Juan be able to make? Use the model.

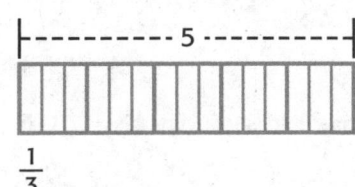

Ⓐ 18 sections   Ⓒ 6 sections

Ⓑ 15 sections   Ⓓ 3 sections

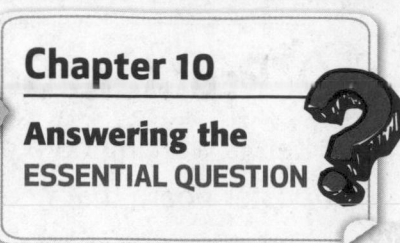
Use what you learned about fraction operations to complete the graphic organizer.

Write the Example

Vocabulary

Model

ESSENTIAL QUESTION

What strategies can be used to multiply and divide fractions?

Real-World Example

Now reflect on the ESSENTIAL QUESTION Write your answer below.

_____

_____

11 Measurement

ESSENTIAL QUESTION

How can I use measurement conversions to solve real-world problems?

My Favorite Animals

Watch

Watch a video!

MY Standards

Measurement and Data

5.MD.1 Convert among different-sized standard measurement units within a given measurement system (e.g., convert 5 cm to 0.05 m), and use these conversions in solving multi-step, real world problems.

5.MD.2 Make a line plot to display a data set of measurements in fractions of a unit $\left(\frac{1}{2}, \frac{1}{4}, \frac{1}{8}\right)$. Use operations on fractions for this grade to solve problems involving information presented in line plots.

I'll be able to get this—no problem!

Standards for
Mathematical
PRACTICE

1. Make sense of problems and persevere in solving them.
2. Reason abstractly and quantitatively.
3. Construct viable arguments and critique the reasoning of others.
4. Model with mathematics.
5. Use appropriate tools strategically.
6. Attend to precision.
7. Look for and make use of structure.
8. Look for and express regularity in repeated reasoning.

= focused on in this chapter

Name

_____

Am I Ready?

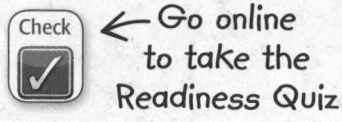 Check ✓ ← Go online to take the Readiness Quiz

Multiply.

1. 12 × 3 = _____

2. 36 × 5 = _____

3. 1,760 × 4 = _____

4. 6 × 1,000 = _____

5. 15 × 100 = _____

6. 947 × 100 = _____

7. A musical was sold out for three straight shows. If 825 tickets were sold at each performance, how many tickets were sold in all?

_____

Divide.

8. 45 ÷ 3 = _____

9. 112 ÷ 16 = _____

10. 39 ÷ 4 = _____

11. 500 ÷ 100 = _____

12. 150 ÷ 10 = _____

13. 7,900 ÷ 100 = _____

14. A box has 144 ounces of grapes. How many 16-ounce packages of grapes can be made?

_____

Shade the boxes to show the problems you answered correctly.

How Did I Do? ➤

1	2	3	4	5	6	7	8	9	10	11	12	13	14

MY Math Words

Vocab
$^a b_c$

Review Vocabulary

capacity   estimate   length   weight

Making Connections

Use the review vocabulary to tell what you would measure
for each question. Then provide estimates for each category.

What is the approximate
distance across a camel's
snout?
_____

About how much
water does a camel's
hump hold?
_____

Provide estimates for each measure.

capacity _____

length _____

weight _____

About how heavy is an adult camel?
_____

MY Foldable

FOLDABLES® Follow the steps on the back to make your Foldable.

1 gallon

4 quarts

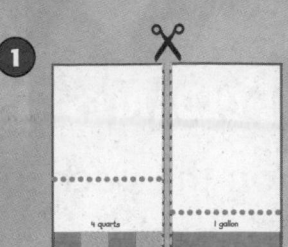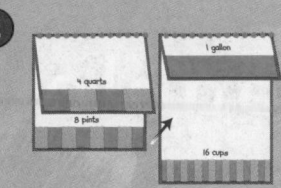 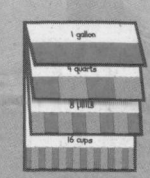
16 cups

8 pints

Hands On
Measure with a Ruler

Lesson 1

ESSENTIAL QUESTION
How can I use measurement conversions to solve real-world problems?

Length is the measurement of distance between two points. You can use a ruler to measure the length of objects to the nearest half inch or quarter inch.

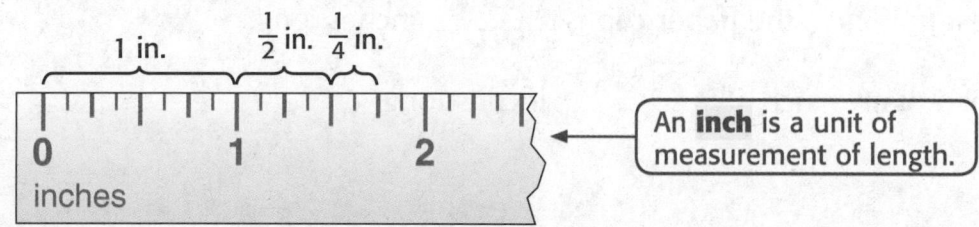

An **inch** is a unit of measurement of length.

If you measure an object to a smaller unit of measure, you get a more precise or accurate measurement.

Measure It

The width of a button is the length of the widest part of the button. Find the width of the button to the nearest half inch and quarter inch.

1 Place the ruler against one edge of the object. Line up the zero on the ruler with the end of the object.

2 Find the half-inch mark that is closest to the other end. Repeat for the quarter-inch mark.

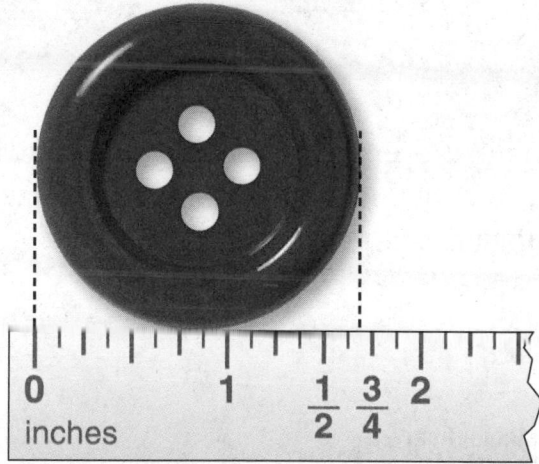

To the nearest half inch, the button is _____ inches wide.

To the nearest quarter inch, it is _____ inches wide.

Cut and use this inch ruler.

Try It

Find the length of a paper clip to the nearest half inch and quarter inch.

1 Place the ruler against one edge of the object. Line up the zero on the ruler with the end of the object.

2 Find the half-inch mark that is closest to the other end. Repeat for the quarter-inch mark.

Helpful Hint

All measurements are approximations. However, if you use smaller units, you will get a more precise measure, or a measure that is closer to the exact measure.

To the nearest half inch, the paper clip is _____ inches long.

To the nearest quarter inch, it is _____ inches long.

Talk About It

1. Explain how you can tell the difference between the half-inch and quarter-inch marks when measuring an object with a ruler.

_____

_____

2. Will you ever have the same answer when measuring to the nearest half inch and measuring to the nearest quarter inch? Explain your reasoning.

_____

3. **Mathematical PRACTICE** **6** **Be Precise** Circle the more precise measurement of the button in the first activity.

$1\frac{1}{2}$ inches

$1\frac{3}{4}$ inches

Practice It

Measure the length of each to the nearest half inch and quarter inch.

4.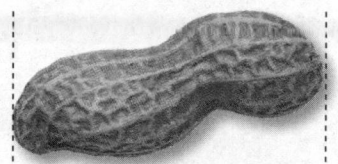

5.

Half Inch: _____

Quarter Inch: _____

Half Inch: _____

Quarter Inch: _____

Find the length of each object to the nearest half inch and quarter inch.

6. width of a book

_____

7. length of a pencil

_____

8. width of a calculator

_____

9. length of a tape dispenser

_____

Draw a line segment with each of the following lengths.

10. $1\frac{1}{4}$ inches

11. $2\frac{1}{2}$ inches

12. $3\frac{3}{4}$ inches

13. The length of a remote controlled car is $8\frac{1}{2}$ inches to the nearest half inch and $8\frac{1}{4}$ inches to the nearest quarter inch. Which measurement is more precise?

_____

_____

Mathematical
14. PRACTICE ⑥ **Be Precise** Carrie and Sam each measured the length of their cat's tail. Carrie measured the length to the nearest half inch. Her measurement was $6\frac{1}{2}$ inches. Sam measured the length to the nearest quarter inch. His measurement was $6\frac{3}{4}$ inches. Circle the correct description of the actual measurement of the cat's tail.

less than $6\frac{1}{2}$ in.   between $6\frac{1}{2}$ in. and $6\frac{3}{4}$ in.   greater than $6\frac{3}{4}$ in.

Mathematical
15. PRACTICE ③ **Find the Error** James used a ruler to measure a piece of string. James said the string is $2\frac{1}{2}$ inches long. Find his mistake and correct it.

```
  0        1        2        3
inches
```

_____

_____

Write About It

16. When might you want to measure the length of an object to the nearest quarter inch as opposed to the nearest half inch?

_____

_____

MY Homework

Homework Helper

Need help?  connectED.mcgraw-hill.com

Find the length of the push pin to the nearest half inch and quarter inch.

1 Place the ruler against one edge of the object. Line up the zero on the ruler with the end of the object.

2 Find the half-inch mark that is closest to the other end. Repeat for the quarter-inch mark.

To the nearest half inch, the push pin is 1 inch long. To the nearest quarter inch, it is $1\frac{1}{4}$ inches long.

Practice

Measure the length of each to the nearest half inch and quarter inch.

1

2.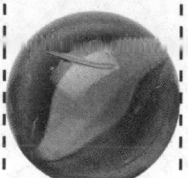

_____   _____

Find the length of each object to the nearest half inch and quarter inch.

3. width of an MP3 player   **4.** length of a stapler

_____   _____

Cut and use this inch ruler.

Draw a line segment with each of the following lengths.

5. $2\frac{1}{4}$ inches

6. $\frac{3}{4}$ inch

Vocabulary Check

7. Choose the correct word(s) to complete the sentence below.

Length is the measurement of _____ between two points.

 # Problem Solving

8. The height of Josie's dog is $11\frac{1}{4}$ inches to the nearest quarter inch and 11 inches to the nearest half inch. Which measurement is more precise?

_____

9. Allison has a ruler that is marked in fourths of an inch and a tape measure that is marked in halves of an inch. Which measuring tool will give Allison a more precise measure?

_____

What's UP?

10. **Mathematical PRACTICE** 6 **Be Precise** Gretchen measured the length of her toaster to the nearest half inch to be 9 inches. Steve measured the toaster to the nearest quarter inch and found that it measured $9\frac{1}{4}$ inches. Who used a more precise measurement?

_____

Convert Customary Units of Length

Lesson 2

ESSENTIAL QUESTION
How can I use measurement conversions to solve real-world problems?

 ## Math in My World Watch Tutor

Example 1

For many of the thrill rides at amusement parks, riders must be at least 48 inches tall. Cheng is 4 feet tall. Is he tall enough to ride a thrill ride at an amusement park?

Convert 4 feet to inches.

Since 1 foot = 12 inches, multiply 4 by 12.

$$\begin{array}{r} 1 \ \ 2 \\ \times \ \ \ 4 \\ \hline \square \ \square \end{array}$$

A A H H H!

So, 4 feet = _____ inches.

Is Cheng tall enough to ride the thrill rides? _____

Check Use division to check your answer. _____ ÷ 4 = 12

The units of length most often used in the United States are the inch, foot, yard, and mile. These units are part of the **customary system.**

Key Concept Customary Units of Length

> 1 **foot** (ft) = 12 **inches** (in.)
> 1 **yard** (yd) = 3 ft or 36 in.
> 1 **mile** (mi) = 5,280 ft or 1,760 yd

When you **convert** measurements, you change from one unit to another. To convert a larger unit to a smaller unit, multiply. To convert a smaller unit to a larger unit, divide.

Online Content at 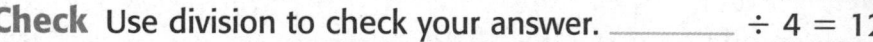 **connectED.mcgraw-hill.com**

Example 2

Convert 42 inches to feet.

Since you are changing a smaller
unit to a larger unit, divide.

$$12\overline{)42}$$ □ R □

Interpret the remainder.

One Way Write the remainder in inches.

The remainder _____ means there are _____ inches left over.

42 inches = _____ feet _____ inches

Another Way Write the remainder as a fraction or decimal.

The fraction of a foot is _____ or _____ .

42 inches = _____ feet

So, 42 inches is equal to _____ feet _____ inches, or _____
feet or 3.5 feet.

Guided Practice

Talk MATH

Explain how to
convert units from
feet to inches.

1. Complete 60 in. = ■ ft

   60 ÷ 12 = _____

   So, 60 inches equals _____ feet.

2. Complete 16 yd = ■ ft

   16 × 3 = _____

   So, 16 yards equals _____ feet.

Independent Practice

Complete.

3. 72 in. = _____ ft

4. 19 yd = _____ in.

5. 40 in. = _____ ft

6. 7,920 ft = _____ mi

7. 6 yd = _____ in.

8. 26,400 ft = _____ mi

9. 9 ft = _____ in.

10. 22 ft = ___ yd ___ ft

11. 51 in. = ___ ft ___ in.

Compare. Use <, >, or = to make a true statement.

12. 24 in. $\bigcirc$ 2 ft 3 in.

13. 7 yd $\bigcirc$ 20 ft

14. 2 mi $\bigcirc$ 3,500 yd

15. $11\frac{1}{2}$ ft $\bigcirc$ 4 yd

16. 72 in. $\bigcirc$ 2 yd

17. 5.5 yd $\bigcirc$ 192 in.

Problem Solving

18. PRACTICE Mathematical **3 Draw a Conclusion** Dana ran $\frac{1}{4}$ mile. Trish ran 445 yards. Who ran the greater distance? Explain.

_____

19. Carlos is 63 inches tall. What is his height in feet?

_____

What is his height in feet and inches?

_____

20. The U.S.S. Harry Truman is an aircraft carrier that is 1,092 feet long. Find the length in yards.

_____

21. A bull had horns that measured 98 inches across. What is this length in feet and inches?

_____

What is this length in feet?

_____

HOT Problems

22. PRACTICE Mathematical **2 Use Number Sense** There are 320 rods in a mile. Find the length of a rod in feet.

_____

23. **Building on the Essential Question** Explain, using an example, why you need to multiply when converting from a larger unit to a smaller unit.

_____

_____

My Work!

Get the point?

MY Homework

Homework Helper

Need help? ⟋ connectED.mcgraw-hill.com

The average height of a female giraffe is 15 feet. What is the average height in yards?

Convert 15 feet to yards.

Since 3 feet = 1 yard, divide 15 by 3.

$$\begin{array}{r} 5 \\ 3\overline{)15} \\ -15 \\ \hline 0 \end{array}$$

So, 15 feet = 5 yards.

The average height of the female giraffe is 5 yards.

Check Use multiplication to check your answer.

$5 \times 3 = 15$

Practice

Complete.

1. 5 mi = _____ yd

2. 7 yd = _____ in.

3. 150 in. = _____ yd

4. 110 in. = ____ ft ____ in.

5. 8 yd = _____ in.

6. 13,200 ft = _____ mi

Vocabulary Check

Fill in the correct circle that corresponds to the best answer.

7. Which of the following is not a common unit of measurement for the customary system?

(A) feet   (C) meter

(B) inches   (D) miles

8. Which operation is necessary to convert a smaller unit to a larger unit?

(F) addition   (H) multiplication

(G) subtraction   (I) division

Problem Solving

Mathematical
9. PRACTICE 6 **Be Precise** Ty has two pieces of wood. Use <, >, or = to compare the pieces of wood.

Piece	Length
1	1 yd 9 in.
2	44 in.

_____

10. Lily built an orange shelf and a brown shelf to hang in her room. The orange shelf is 5 feet 6 inches long. The brown shelf is twice as long. What is the length of the brown shelf?

_____

How tall am I?

Test Practice

11. The table shows the name of each volunteer and the height of the baby elephant each volunteer measured. Which volunteer measured the tallest elephant?

Elephant Heights	
Volunteer	**Height of Elephant**
George	2 yd
Kelsey	68 in.
Mariah	5 ft
Bryan	5 ft 10 in.

(A) George   (C) Mariah

(B) Kelsey   (D) Bryan

Problem-Solving Investigation

STRATEGY: Use Logical Reasoning

Lesson 3

ESSENTIAL QUESTION
How can I use measurement conversions to solve real-world problems?

Learn the Strategy

Three friends each measured their height. Their heights are 4 feet 10 inches, 4 feet 9 inches, and 4 feet 7 inches. Use the clues to determine the height, in inches, of each person.
- Elliot is taller than Jorge.
- Nicole is 3 inches taller than the shortest person.
- Elliot is 57 inches tall.

1 Understand

What facts do you know?

the clues that are listed above

What do you need to find?

the _____ of each person

2 Plan

I can use logical reasoning to find the height of each person.

3 Solve

Convert the measurements to inches to compare.

$4 \times 12 + 10 = 48 + 10 =$ _____ inches   Nicole is _____ inches tall.

$4 \times 12 + 9 = 48 + 9 =$ _____ inches   Elliot is _____ inches tall.

$4 \times 12 + 7 = 48 + 7 =$ _____ inches   Jorge is _____ inches tall.

4 Check

Is my answer reasonable?

Since all of the answers match the clues, the solution is reasonable.

Practice the Strategy

Three dogs are sitting in a line. Rocky is not last. Coco is in front of the tallest dog. Marley is sitting directly behind Rocky. List the dogs in order from first to last.

Who am I?

1 Understand

What facts do you know?

_____

_____

_____

What do you need to find?

_____

2 Plan

_____

_____

3 Solve

4 Check

Is my answer reasonable?

_____

Apply the Strategy

Solve each problem by using logical reasoning.

1. An after-school club is building a clubhouse that has a rectangular floor that is 8 feet by 6 feet. What is the total floor area, in square inches, of the clubhouse?

2. There is a red, a green, and a yellow bulletin board hanging in the hallway. All of the bulletin boards are rectangular with a height of 4 feet. Their lengths are 6 feet, 5 feet, and 3 feet. The red bulletin board has the largest area and the yellow one has the smallest area. What is the area of the green bulletin board?

3. **Mathematical PRACTICE 8 Look for a Pattern** If the pattern below continues, how many pennies will be in the fifth figure?

Figure 1   Figure 2   Figure 3

4. A cafeteria table has an area of 21 square feet. If three tables are pushed together, what is the combined area of the tables?

5. Ethan has $1.25 in dimes, nickels, and pennies. He has twice as many dimes as pennies, and the number of nickels is one less than the number of pennies. How many dimes, nickels, and pennies does he have?

Review the Strategies

Use any strategy to solve each problem.

- Use logical reasoning.
- Draw a diagram.
- Look for a pattern.
- Solve a simpler problem.

6. Madeline has 2 times the number of games as Paulo. Paulo has 4 more games than Tyler. If Tyler has 9 games, how many games are there among the 3 friends?

_____

7. When Cheryl goes mountain climbing, she rests 5 minutes for every 15 minutes that she climbs. If Cheryl's combined time is 2 hours, how many minutes does she rest?

_____

8. There are 8 girls for every 7 boys on a field trip. If there are 56 girls on the trip, how many students are on the trip?

_____

9. There are 4 more girls in Mrs. Pitt's class than Mr. Brown's class. Five girls moved from Mrs. Pitt's class to Mr. Brown's class. Now there are twice as many girls in Mr. Brown's class as there are in Mrs. Pitt's. How many girls were in Mr. Brown's class to begin with?

_____

10. A storage room measures 48 inches by 60 inches. What is the total area, in square feet, of the closet?

_____

11. Mathematical **PRACTICE** 1 **Make Sense of Problems** Five friends go to a batting cage. Andrea bats after Daniel and before Jessica. Juwan bats after Andrea and before Jessica and Filipe. Jessica always bats immediately after Juwan. Who bats last?

_____

My Work!

Batter up!

MY Homework

Homework Helper

Need help? connectED.mcgraw-hill.com

A family has three guinea pigs. Tiger is 8 years old, which is 2 years older than Max. Max is 3 years older than Patches. List the guinea pigs from oldest to youngest.

1 Understand

What facts do you know?
Tiger is 8 years old.
Tiger is 2 years older than Max, and Max is 3 years older than Patches.

What do you need to find?
the ages of the guinea pigs from oldest to youngest

2 Plan

Use logical reasoning to find the age of each guinea pig.
Make a table to help organize the information.

3 Solve

Place an "X" in each box that cannot be true.

You know that Tiger is 8 years old.
Subtract 2 from Tiger's age to find Max's age. Max is 6 years old.
Subtract 3 from Max's age to find Patches' age. Patches is 3 years old.

	oldest	2nd oldest	youngest
Tiger	yes	X	X
Max	X	yes	X
Patches	X	X	yes

So, Tiger is the oldest, Max is the next oldest, and the youngest is Patches.

4 Check

Is my answer reasonable?
Since all of the answers match the clues, the solution is reasonable.

Problem Solving

Solve each problem by using logical reasoning.

1. **Mathematical PRACTICE 5 Use Math Tools**
   Mr. Toshi's fifth grade class sold containers of popcorn and peanuts. If each day they sold 25 fewer containers of peanuts than popcorn, how many containers of popcorn and peanuts did they sell in all? Complete the table and solve.

	Day 1	Day 2	Day 3	Day 4
Popcorn	225	200	150	300
Peanuts				

   _____

2. A shelter house has a floor area of 400 square feet. If three identical shelter houses are built, what is the combined floor area of the shelter houses?

   _____

3. Meghan has $1.10 in dimes, nickels, and pennies. She has three times as many nickels as pennies, and the number of dimes is two less than the number of pennies. How many dimes, nickels, and pennies does she have?

   _____

4. Alana is 4 years older than her brother Ernie. Ernie is 2 years older than their sister Amelia. Amelia is 10 years younger than their brother Mazo. If Mazo is 17 years old, how old is Alana?

   _____

I dig carrots!

5. A rectangular garden measures 15 feet by 30 feet. What is the total area, in square yards, of the garden?

   _____

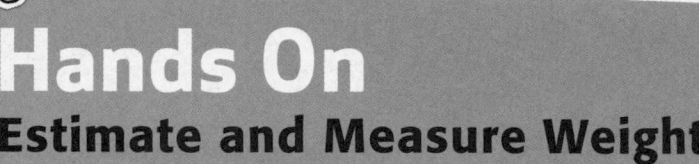

Hands On
Estimate and Measure Weight

Lesson 4

ESSENTIAL QUESTION
How can I use measurement conversions to solve real-world problems?

The **weight** of an object is a measure of how heavy it is. **Ounces** and **pounds** are examples of units of measurement for weight.

Measure It

1 Estimate the weight of each object in ounces or pounds. Record your results in the table.

Object	Estimate	Actual
Eraser		
Glue bottle		
Math book		
Pencil		

2 Measure the weight of each object. Place the eraser on one side of a balance. Set ounce or pound weights on the other side until the sides are level. Record the actual weight. Repeat this step for the other objects.

Try It

1 Place a 1-pound weight on one side of a balance.

2 Place 1-ounce weights on the opposite side of the balance. Continue placing 1-ounce weights until the balance is equal. Use your results to complete the table.

Pounds (lb)	Ounces (oz)
1	
2	
3	
4	

So, a 1-pound weight is equal to _____ ounces.

Talk About It

1. Order the four objects you weighed in the activity from greatest to least weight.

_____

Mathematical
2. PRACTICE 6 **Explain to a Friend** Is the total weight of the four objects you measured in the first activity greater than 2 pounds? Explain.

_____

_____

_____

3. Write a sentence that describes the relationship that is usually found between an object's size and weight.

_____

_____

Practice It

4. Identify three objects in your classroom that you can use a balance to find their weights. Estimate each object's weight. Then weigh each object and record the actual weight in the table below.

Object	Estimate	Actual

Compare. Use >, <, or = to make a true statement.

5. 46 ounces ◯ 3 pounds

6. 96 ounces ◯ 6 pounds

7. 130 ounces ◯ 8 pounds

8. 113 ounces ◯ 7 pounds

9. 5 pounds ◯ 78 ounces

10. 9 pounds ◯ 145 ounces

11. 10 pounds ◯ 160 ounces

12. 12 pounds ◯ 196 ounces

My Work!

13. Tim and Cameron measured the weight of the same calculator. Tim measured the calculator as 1 pound. Cameron measured the calculator as 13 ounces. Circle the greater measurement.

      1 pound   13 ounces

14. Cara measured a dictionary twice. Her first measurement was 1 pound. The second measurement was 28 ounces. Circle the greater measurement.

      1 pound   28 ounces

15. If you are measuring a bag of sugar, would ounces or pounds give you a more precise measurement? Explain.

_____

_____

Pouncing ounces!

16. **Mathematical PRACTICE 6 Be Precise** Jeffrey measured the weight of his kitten. His first measurement was 34 ounces. His second measurement was 2 pounds. Use >, <, or = to make a true statement.

      34 ounces ◯ 2 pounds

17. **Mathematical PRACTICE 3 Draw a Conclusion** Compare and contrast pounds and ounces.

_____

_____

Write About It

18. Why is estimation of weight important in everyday life?

_____

_____

MY Homework

Homework Helper

Need help? connectED.mcgraw-hill.com

Sixteen 1-ounce weights are equal to 1 pound. Use the table below to determine a pattern. Then find the weight, in ounces, that is equal to 4 pounds.

Pounds (lb)	Ounces (oz)	
1	16	
2	32	+16
3	48	+16
4	64	+16

So, a 4-pound weight is equal to 48 + 16, or 64 ounces.

Practice

Compare. Use >, <, or = to make a true statement.

1. 30 ounces ◯ 2 pounds

2. 98 ounces ◯ 6 pounds

3. 11 pounds ◯ 176 ounces

4. 7 pounds ◯ 109 ounces

Vocabulary Check

5. Fill in the blank with the correct word to complete the sentence.

Weight is a measure of how _____ an object is.

Problem Solving

6. Kurt measured the weight of his ball python snake. His first measurement was 48 ounces. His second measurement was 3 pounds. Compare the two measurements. Use >, <, or = to make a true statement.

_____

7. Lori and Mindy measured the weight of their book bags. Lori measured her bag using pounds. Mindy measured her bag using ounces. Circle the measure that is more appropriate to measure the weight of the book bag.

pounds   ounces

8. Tracy measured the weight of her hamster. Her first measurement was 13 ounces. Her second measurement was 1 pound. Which measurement is more precise?

_____

Mathematical
9. PRACTICE 6 Be Precise Quinton measured the weight of his laptop. His first measurement was 94 ounces. His second measurement was 6 pounds. Compare the two measurements. Use >, <, or = to make a true statement.

_____

10. If you are selling bags of candy, would you want to sell the candy in ounces or pounds? Explain.

_____

_____

Convert Customary Units of Weight

Lesson 5

ESSENTIAL QUESTION
How can I use measurement conversions to solve real-world problems?

Weight is a measure of how heavy an object is. Customary units of weight are ounce, pound, and ton.

 **Math in My World** Watch Tutor

A kitten at heart!

Example 1

A newborn lion cub weighs about 5 pounds. How many ounces does the cub weigh?

Convert 5 pounds to ounces.

Since 1 pound = 16 ounces, multiply 5 by 16.

So, 5 pounds = _____ ounces.

A newborn lion cub weighs about _____ ounces.

Check Use division to check your answer.

_____ ÷ 5 = 16

$$\begin{array}{r} 1\ \ 6 \\ \times\quad 5 \\ \hline \square\ \square \end{array}$$

Key Concept   Customary Units of Weight

1 pound (lb) = 16 **ounces** (oz)   **1 ton** (T) = 2,000 pounds

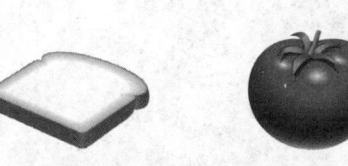

1 ounce   1 pound   1 ton

As with units of length, units of weight in the customary system can also be expressed using different units or as fractions and decimals.

Example 2

Convert 56 ounces to pounds.

Since you are changing a smaller unit to a larger unit, divide.

The remainder _____ means there are _____ ounces left over.

$$16\overline{)5\quad6}\quad\square\ R\ \square$$

The fraction of a pound is $\frac{8}{16}$ or $\frac{1}{2}$.

So, 56 ounces = _____ pounds _____ ounces, $3\frac{1}{2}$ pounds, or 3.5 pounds.

Talk MATH

Explain how to compare 22 ounces to 2 pounds.

Guided Practice

Complete.

1. 3 lb = ■ oz

   $3 \times 16 =$ _____

   So, 3 pounds equals _____ ounces.

2. $2\frac{1}{2}$ lb = ■ oz

   $2\frac{1}{2} \times 16 =$ _____

   So, $2\frac{1}{2}$ pounds equals _____ ounces.

3. 32 oz = ■ lb

   $32 \div 16 =$ _____

   So, 32 ounces equals _____ pounds.

Independent Practice

Complete.

4. 96 oz = _____ lb

5. 7 T = _____ lb

6. 1.5 T = _____ lb

7. 11,000 lb = _____ T

8. 50 oz = _____ lb _____ oz

9. $1\frac{1}{4}$ lb = _____ oz

10. 7,000 lb = _____ T

11. $\frac{3}{4}$ T = _____ lb

Compare. Use <, >, or = to make a true statement.

12. 7,500 lb ◯ 4 T

13. $\frac{1}{2}$ T ◯ 1,000 lb

14. 1,200 oz ◯ 72 lb

15. 6 lb 11 oz ◯ 117 oz

 **Problem Solving**

16. Mia combines the items in the table to make potting soil. Order the items according to weight, from least to greatest.

Item	Amount
Topsoil	3 lb
Fertilizer	2 lb 9 oz
Bone meal	43 oz

_____

17. How many 4-ounce bags of peanuts can be filled from a 5-pound bag of peanuts?

_____

18. Alfonso mails a package that weighs 9 pounds. How many ounces is this package?

_____

Mathematical
19. PRACTICE ➋ **Use Number Sense** A puppy weighs 12 ounces. What fractional part of a pound is this?

_____

HOT Problems

Mathematical
20. PRACTICE ➋ **Use Number Sense** A baby weighs 8 pounds 10 ounces. If her weight doubles in 6 months, how much will she weigh in 6 months?

_____

21. ❓ **Building on the Essential Question** How can I apply converting customary units of weight in everyday life?

_____

_____

_____

MY Homework

Lesson 5

Convert Customary Units of Weight

Homework Helper

Need help? connectED.mcgraw-hill.com

A white rhinoceros weighs about $2\frac{1}{2}$ tons. How many pounds does the white rhinoceros weigh?

Convert $2\frac{1}{2}$ tons to pounds.

You know that 2 tons = 4,000 pounds and $\frac{1}{2}$ ton = 1,000 pounds.

So, $2\frac{1}{2}$ tons = 4,000 + 1,000 or 5,000 pounds.

A white rhinoceros weighs about 5,000 pounds.

Practice

Complete.

1. 64 oz = _____ lb

2. 5 T = _____ lb

3. 13 lb = _____ oz

4. 9,000 lb = _____ T

5. 45 oz = _____ lb _____ oz

6. 3.25 T = _____ lb

Problem Solving

7. A hippopotamus weighs 1.5 tons. What is its weight in ounces?

_____

I'm a TON of fun!

8. A restaurant serves a 20-ounce steak. How much does the steak weigh in pounds and ounces?

_____

9. **Mathematical**
PRACTICE 2 **Use Number Sense** What fractional part of a pound is an ounce?

_____

Vocabulary Check

Circle the correct term to complete each sentence.

10. (Ounces, Pounds) would be the most appropriate customary unit to measure the weight of a paper clip.

11. (Weight, Length) is a measure of how heavy an object is.

12. (Tons, Pounds) would be the most appropriate customary unit to measure the weight of a bag of flour.

Test Practice

13. A rabbit weighs 4 pounds 6 ounces. How many ounces does the rabbit weigh?

> 1 pound = 16 ounces

Ⓐ 70 oz   Ⓒ 16 oz

Ⓑ 64 oz   Ⓓ 6 oz

Check My Progress

Vocabulary Check

Choose the correct word that completes each sentence.

foot	inches	length	mile
pound	ounce	ton	yard

1. The _____ is an appropriate unit to measure the distance ran over a one-week period.

2. The _____ is an appropriate unit to measure the weight of a bag of sugar.

3. The _____ is an appropriate unit to measure the distance ran on a football field.

Concept Check

Draw a line segment with each of the following lengths.

4. $2\frac{3}{4}$ inches

5. $\frac{1}{2}$ inch

Complete.

6. 72 in. = _____ ft

7. 11,880 ft = _____ mi

8. 3.75 yd = _____ in.

9. 17 ft = _____ yd _____ ft

Complete.

10. 124 oz = _____ lb

11. 8 T = _____ lb

12. 7 T 500 lb = _____ lb

13. $1\frac{1}{2}$ lb = _____ oz

Compare. Use >, <, or = to make a true statement.

14. 74 ounces $\bigcirc$ 4 pounds

15. 5 pounds $\bigcirc$ 81 ounces

Problem Solving

16. An African lion weighs 450 pounds. What is its weight in ounces?

_____

17. Lin and Pazi each think of a number. Lin's number is 7 more than Pazi's number. The sum of the two numbers is 49. What is Pazi's number?

_____

18. Janna jumped 156 inches in the long jump competition at the high school track meet. How many feet did Janna jump?

_____

Test Practice

19. How many 8-ounce bags of trail mix can be filled from a 10-pound bag of trail mix?

   Ⓐ 5 bags   Ⓒ 20 bags

   Ⓑ 10 bags   Ⓓ 25 bags

Hands On
Estimate and Measure Capacity

Lesson 6

ESSENTIAL QUESTION
How can I use measurement conversions to solve real-world problems?

Capacity is the measure of how much a container can hold. **Cups**, **pints**, and **gallons** are examples of units of measurement for capacity.

Measure It

1. Estimate the capacity of a pint container. Record your estimate in the table.

Container	Estimate	Actual
Pint	_____ cups	_____ cups
Quart	_____ pints	_____ pints

2. Fill one cup with water and pour its contents into the pint container. Repeat until the pint container is full.

   How many cups does it take to fill the

   pint container? _____

   So, there are _____ cups in a pint.

   Record the measure in the table.

3. Estimate the capacity of a quart container. Record your estimate in the table.

4. Fill one pint with water and pour its contents into the quart container. Repeat until the quart container is full.

   How many pints does it take to fill the

   quart container? _____

   So, there are _____ pints in a quart.

   Record the measure in the table.

Try It

1 Estimate the capacity of a gallon container. Record your estimate in the table.

Container	Number of Quarts	
	Estimate	Actual
Gallon	_____ quarts	_____ quarts

2 Fill one quart container with water and pour its contents into the gallon container. Repeat until the gallon container is full.

How many quarts does it take to fill the gallon container?

_____

So, there are _____ quarts in a gallon.

Record the measure in the table.

Talk About It

1. Is the capacity of the gallon container less than or greater than the capacity of the pint container? Explain.

_____

_____

Mathematical
2. PRACTICE 4 Model Math Use the pint container to estimate and measure how many pints are in a gallon. Record your results in the table below.

Container	Number of Pints	
	Estimate	Actual
Gallon	_____ pints	_____ pints

3. Explain how you could convert 8 quarts to gallons without measuring.

_____

_____

Practice It

Complete using your findings from the activities.

4. 3 pints = _____ cups   **5.** 4 gallons = _____ pints

6. 12 cups = _____ pints   **7.** 3 gallons = _____ quarts

8. 24 quarts = _____ gallons   **9.** 20 cups = _____ quarts

10. 3 quarts = _____ cups   **11.** 2 gallons = _____ pints

Compare. Use >, <, or = to make a true statement.

12. 8 pints ◯ 15 cups   **13.** 12 cups ◯ 8 pints

14. 4 gallons ◯ 30 pints   **15.** 2 quarts ◯ 8 cups

My Work!

16. Seth made 4 quarts of fruit punch for a birthday party. How many cups of fruit punch did he make?

_____

17. If you measure the amount of water you drink in a day, would cups or pints give you a more precise measurement? Explain.

_____

_____

_____

Mathematical
18. PRACTICE **6** **Be Precise** Peggy measured the capacity of her aquarium. Her first measurement was 9 gallons. Her second measurement was 35 quarts. Compare the two measurements. Use >, <, or = to make a true statement.

_____

Mathematical
19. PRACTICE **3** **Justify Conclusions** Is it faster to water two large flower pots using a one-cup pitcher or a one-quart pitcher? Explain.

_____

_____

Write About It

20. How can I use measuring tools to find capacity?

_____

_____

MY Homework

Lesson 6

Hands On: Estimate and Measure Capacity

Homework Helper

Need help? ⟋ connectED.mcgraw-hill.com

The table shows the measures you found in the activities.

How many quarts equal 10 pints?

Since 2 pints = 1 quart, divide by 10 by 2.

$10 \div 2 = 5$

So, 5 quarts is equal to 10 pints.

Customary Units of Capacity
1 pint = 2 cups
1 quart = 2 pints
1 gallon = 4 quarts

Practice

Complete.

1. 4 cups = _____ pints

2. 16 quarts = _____ gallons

3. 8 quarts = _____ pints

4. 7 pints = _____ cups

5. 5 gallons = _____ pints

6. 28 cups = _____ quarts

Vocabulary Check

7. Fill in the blank with the correct word to complete the sentence below.

_____ is the measure of how much a container can hold.

8. Mathematical PRACTICE 2 Reason Sophie drank 8 cups of water one day. How many pints of water did she drink?

_____

9. A container holds 26 pints of water. Circle whether 26 pints is _greater than, less than,_ or _equal to_ 12 quarts.

greater than   less than   equal to

10. Coach Cole measured the capacity of his water cooler. His first measurement was 6 gallons. His second measurement was 25 quarts. Compare the two measurements. Use >, <, or = to make a true statement.

_____

11. If you are measuring the amount of water evaporated from a swimming pool, would gallons or quarts give you a more precise measurement? Explain.

_____

_____

_____

12. Jason is a zookeeper and needs to fill a large tub of water for a zebra. Would it be faster to use a one-quart container or a one-gallon container? Explain.

SLURP!

6 gallons

_____

_____

Convert Customary Units of Capacity

Lesson 7

ESSENTIAL QUESTION
How can I use measurement conversions to solve real-world problems?

Capacity is the measure of how much a container can hold. Just like length and weight, the smaller the unit you use to measure capacity, the more precise the measurement.

 ## Math in My World

Example 1

Sondra tries to drink 9 cups of water each day. How many fluid ounces of water is 9 cups?

Convert 9 cups to fluid ounces.

Since 1 cup = 8 fluid ounces, multiply 9 by 8.

$$\begin{array}{r} 9 \\ \times\ 8 \\ \hline \square\ \square \end{array}$$

So, 9 cups = _____ fluid ounces.

Key Concept Customary Units of Capacity

1 **cup** (c) = 8 **fluid ounces** (fl oz)
1 **pint** (pt) = 2 c = 16 fl oz
1 **quart** (qt) = 2 pt = 32 fl oz
1 **gallon** (gal) = 4 qt = 128 fl oz

Helpful Hint
A fluid ounce is different from the ounce used to measure weight.

 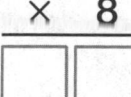  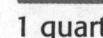 

1 fl oz   1 cup   1 pint   1 quart   1 gallon

Online Content at ⟋ **connectED.mcgraw-hill.com**

As with units of length and units of weight, measurements of capacity in the customary system can also be expressed as fractions.

Example 2

How many quarts can be made from 7 pints?

Since 2 pints = _____ quart, divide 7 by _____ .

The remainder _____ means there is _____ pint left over.

The fraction of a quart is _____ .

So, 7 pints = _____ quarts _____ pint, $3\frac{1}{2}$ quarts, or 3.5 quarts.

$$\begin{array}{r} \boxed{}\ R\ \boxed{} \\ 2\overline{)\ 7} \\ -\boxed{} \\ \hline \boxed{} \end{array}$$

Guided Practice

Complete.

1. 3 c = ■ fl oz

3 × 8 = _____

So, 3 cups equals _____ fluid ounces.

2. 4 qt = ■ c

4 × 2 = _____ pt

8 × 2 = _____ c

So, 4 quarts equals _____ cups.

3. 18 pt = ■ qt

18 ÷ 2 = _____

So, 18 pints equals _____ quarts.

Talk MATH

Explain how to compare 18 fluid ounces to 2 pints.

Independent Practice

Complete.

4. 5 c = _____ fl oz   **5.** 19 qt = _____ gal _____ qt

6. 50 c = _____ pt   **7.** 22 fl oz = _____ c _____ fl oz

8. 2 gal = _____ fl oz   **9.** 7 c = _____ pt

10. 19 c = _____ fl oz   **11.** 25 gal = _____ qt

12. 68 pt = _____ gal   **13.** 56 qt = _____ gal

Compare. Use <, >, or = to make a true statement.

14. 4 c ◯ 40 fl oz   **15.** 1.75 gal ◯ 10 pt   **16.** $2\frac{3}{4}$ qt ◯ $5\frac{1}{2}$ pt

17. 3 gal ◯ 13 qt   **18.** 5 qt ◯ 10 pt   **19.** 300 fl oz ◯ 2 gal

Problem Solving

My Work!

20. Mathematical PRACTICE ⑥ **Explain to a Friend** A bucket contains $1\frac{1}{2}$ gallons of water. Is the amount of water in the bucket *greater than*, *less than*, or *equal to* 6 quarts? Explain.

_____

21. Zach had 1 quart of milk. He used 1 pint to make pancakes and 1 cup to make scrambled eggs. How many cups of milk were left?

_____

22. A cow can drink up to 35 gallons of water in a day. How many quarts of water can a cow drink in a day?

_____

HOT Problems

23. Mathematical PRACTICE ⑥ **Be Precise** Trevor needs to use the most precise measurement when measuring orange juice to add to a mix. Which unit of measure should he use?

_____

24. ❓ **Building on the Essential Question** How can I convert units of capacity?

_____

_____

_____

I'm a quart low!

MY Homework

Homework Helper

Need help? ⤢ connectED.mcgraw-hill.com

Paco has 3 pints of juice plus 1 cup of juice. How many cups of juice does he have in all?

Convert 3 pints to cups.

Since 1 pint = 2 cups, multiply 3 by 2.

So, 3 pints = 6 cups.

$$\begin{array}{r} 3 \\ \times\ 2 \\ \hline 6 \end{array}$$

Then add the remaining 1 cup.   6 + 1 = 7

So, Paco has 7 cups of juice.

Practice

Complete.

1. 16 c = _____ pt

2. 64 fl oz = _____ c

3. 5 qt = _____ gal _____ qt

4. 18 qt = _____ gal

5. 7 c = _____ pt

6. 16 pt 1 c = _____ c

Problem Solving

7. The table shows the amount of paint left in each jar. Which jar contains the greatest amount of paint? the least?

Jar	Amount
blue	2 pt 4 fl oz
purple	5 cups
green	39 fl oz

_____

Mathematical
8. PRACTICE  **Model Math** Write about a real-world situation that can be solved by converting customary units of capacity. Then solve.

_____

_____

Vocabulary Check

Circle the correct term to complete each sentence.

9. (Cups, Gallons) would be the most appropriate customary unit to measure the capacity of a hot chocolate mug.

10. (Gallons, Pints) would be the most appropriate customary unit to measure the capacity of a swimming pool.

11. (Weight, Capacity) is a measure of how much a container can hold.

Test Practice

12. The average person drinks 1 pint of milk a day. At this rate, how many gallons will a person drink in a leap year (366 days)?

  (A) 45 gallons   (C) $45\frac{3}{4}$ gallons

  (B) $45\frac{1}{2}$ gallons   (D) 46 gallons

Display Measurement Data on a Line Plot

Lesson 8

ESSENTIAL QUESTION
How can I use measurement conversions to solve real-world problems?

 **Math in My World**  Tutor

Example 1

Six friends shared several foot-long submarine sandwiches. The table shows the amount each friend ate. Make a line plot of the lengths in the table.

Sandwich Lengths (ft)		
$\frac{1}{3}$	$\frac{1}{3}$	$\frac{1}{3}$
$\frac{1}{2}$	$\frac{1}{4}$	$\frac{1}{4}$

 Count the number of times each fraction appears in the table.

$\frac{1}{4}$ appears _____ times.

$\frac{1}{3}$ appears _____ times.

$\frac{1}{2}$ appears _____ time.

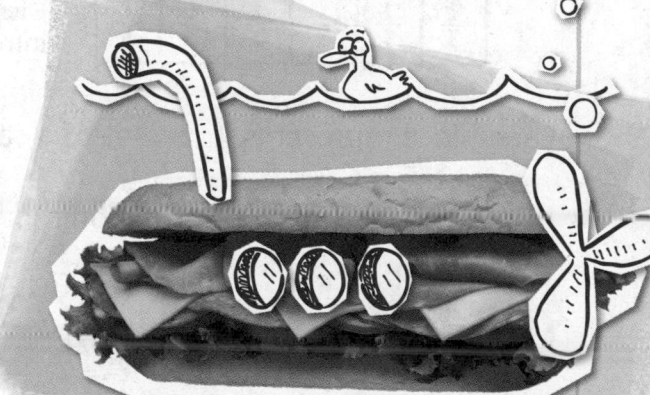

2 Place the correct number of Xs above each fraction on the number line.

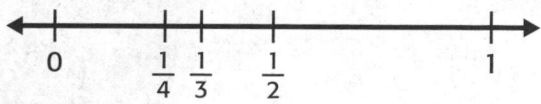

3 Add a title to the line plot.

You can find the **fair share**, or the amount each friend would receive if the sandwiches were divided equally. First add the measurements to find the whole. Then divide the whole by the number of measurements.

Example 2

Find the fair share using the line plot from Example 1.

 Add the fractions to find the total amount of sandwiches eaten. Add the fractions with like denominators first.

So, $\frac{1}{2} + 1 + \frac{1}{2}$ or _____ whole sandwiches were eaten.

2 Xs above $\frac{1}{4}$: $\frac{1}{4} + \frac{1}{4} = \frac{2}{4}$ or $\frac{1}{2}$

3 Xs above $\frac{1}{3}$: $\frac{1}{3} + \frac{1}{3} + \frac{1}{3} = \frac{3}{3}$ or 1

1 X above $\frac{1}{2}$: $\frac{1}{2}$

 Divide the total amount by the number of Xs on the line plot. To find $2 \div 6$, you can draw a model.

2 sandwiches

1	1

$\frac{1}{3}$	$\frac{1}{3}$	$\frac{1}{3}$	$\frac{1}{3}$	$\frac{1}{3}$	$\frac{1}{3}$

Draw 2 rectangles to show 2 whole sandwiches.

Divide the entire amount into 6 equal pieces.

Each piece represents _____ of a sandwich. So, if the sandwiches were

divided equally, each person would have eaten _____ of a sandwich.

Guided Practice

1. Make a line plot of the measurements in the table. Then find the fair share.

Amount of Juice (gal)							
$\frac{1}{4}$	$\frac{1}{4}$	$\frac{1}{4}$	$\frac{1}{2}$	$\frac{1}{2}$	$\frac{1}{2}$	$\frac{1}{4}$	$\frac{1}{2}$

Talk MATH

Describe a situation in everyday life in which you would want to find a fair share.

fair share: _____

Independent Practice

Make a line plot of the measurements in each table. Then find the fair share.

2.

Yarn Lengths (ft)					
$\frac{1}{4}$	$\frac{1}{2}$	$\frac{1}{3}$	$\frac{1}{3}$	$\frac{1}{4}$	$\frac{1}{3}$
$\frac{1}{4}$	$\frac{3}{4}$	$\frac{1}{2}$	$\frac{1}{2}$	$\frac{3}{4}$	$\frac{1}{4}$

fair share: _____

3.

Iced Tea (qt)								
$\frac{1}{2}$	$\frac{1}{4}$	$\frac{1}{8}$	$\frac{1}{4}$	$\frac{1}{2}$	$\frac{1}{4}$	$\frac{1}{2}$	$\frac{1}{8}$	$\frac{1}{2}$

fair share: _____

4.

Amount of Sliced Turkey (lb)							
$\frac{1}{8}$	$\frac{1}{4}$	$\frac{1}{8}$	$\frac{1}{4}$	$\frac{1}{2}$	$\frac{1}{2}$	$\frac{1}{8}$	$\frac{1}{8}$
$\frac{7}{8}$	$\frac{1}{4}$	$\frac{3}{4}$	$\frac{3}{8}$	$\frac{1}{4}$	$\frac{3}{8}$	$\frac{1}{2}$	$\frac{5}{8}$

fair share: _____

5.

Distance Swam (mi)				
$\frac{1}{2}$	$\frac{1}{2}$	$\frac{1}{3}$	$\frac{1}{3}$	$\frac{1}{4}$
$\frac{1}{3}$	$\frac{1}{4}$	$\frac{1}{2}$	$\frac{1}{2}$	$\frac{1}{2}$

fair share: _____

Problem Solving

For Exercises 6–7, use the line plot that shows the amount of rainfall over a twelve-month period in the city of Middleton.

Middleton Rainfall (in.)

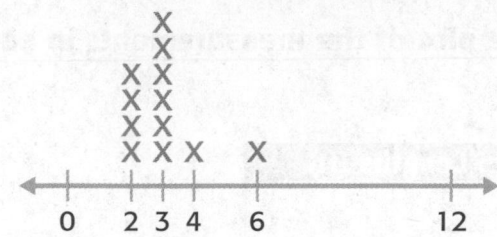

6. **Mathematical** **PRACTICE** **4** **Model Math**
   Convert the rainfall amounts to feet and make a new line plot.

My Drawing!

RAIN, RAIN, GO AWAY!

7. What is the fair share, in feet, if the same amount of rain fell each month?

_____

HOT Problems

8. **Mathematical** **PRACTICE** **2** **Use Number Sense** A unit fraction is a fraction with a numerator of 1. What is the greatest unit fraction that you can plot on a number line between 0 and 1? Explain.

_____

_____

9. **?** **Building on the Essential Question** How can I find the fair share of a set of measurements?

_____

_____

_____

MY Homework

Homework Helper

Need help? connectED.mcgraw-hill.com

The zoo lists the weights of several animals in the table. Make a line plot of the weights in the table.

Animal Weights (T)				
$\frac{1}{8}$	$\frac{1}{2}$	$\frac{1}{8}$	$\frac{1}{8}$	$\frac{1}{2}$
$\frac{1}{8}$	$\frac{1}{2}$	$\frac{1}{2}$	$\frac{1}{4}$	$\frac{1}{4}$

1 Count the number of times each fraction appears in the table.

$\frac{1}{8}$ appears 4 times.

$\frac{1}{4}$ appears 2 times.

$\frac{1}{2}$ appears 4 times.

2 Place the correct number of Xs above each fraction on the number line.

3 Then, use the title from the table to add a title to the line plot.

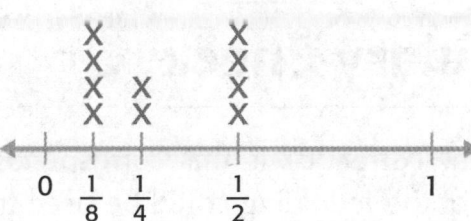
Animal Weights (T)

Practice

Refer to the Homework Helper to answer Exercises 1 and 2.

1. Which weight(s) occurred the most often?

_____

2. Find the fair share.

_____

Problem Solving

3. Make a line plot of the measurements in the table.

Amount of Cashews (lb)					
$\frac{1}{2}$	$\frac{1}{4}$	$\frac{1}{4}$	$\frac{3}{4}$	$\frac{1}{2}$	$\frac{3}{4}$
$\frac{1}{2}$	$\frac{1}{2}$	$\frac{3}{4}$	$\frac{1}{4}$	$\frac{1}{4}$	$\frac{3}{4}$

We're nuts about Math!

4. Refer to the table in Exercise 3. What is the fair share, in pounds, of the cashews?

_____

Mathematical
5. PRACTICE  **Use Math Tools** Explain how you can mentally find the sum of the fractions in Exercise 3 without actually performing the calculations.

_____

_____

_____

Vocabulary Check

6. Fill in the correct circle that corresponds to the best answer. Which of the following could be used to find the amount each person would receive if it was divided equally?

Ⓐ fair share   Ⓒ length

Ⓑ customary system   Ⓓ weight

Test Practice

7. Which is the correct fair share for the measurements shown in the line plot?

Ⓐ $\frac{1}{6}$ mile   Ⓒ $\frac{1}{2}$ mile

Ⓑ $\frac{1}{3}$ mile   Ⓓ $\frac{2}{3}$ mile

Hiking Distance (mi)

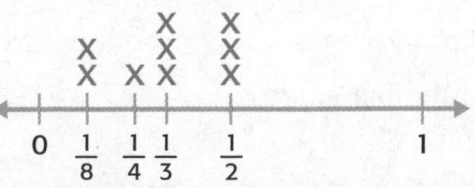

Need more practice? Download Extra Practice at ⚡**connectED.mcgraw-hill.com**

Measurement and Data

5.MD.1

Hands On
Metric Rulers

Lesson 9

ESSENTIAL QUESTION
How can I use measurement conversions to solve real-world problems?

Use a ruler like the one shown to measure objects to the nearest centimeter or to the nearest millimeter.

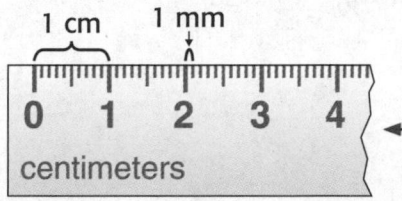

Centimeters and **millimeters** are units of length.

1 centimeter = 10 millimeters

Measure It Tools

Find the length of the piece of chalk to the nearest centimeter.

1 Place the ruler against the piece of chalk. Line up the zero on the ruler with the end of the piece of chalk.

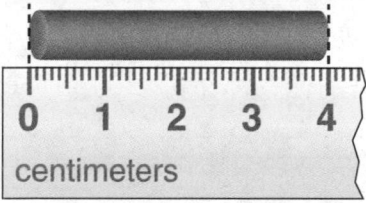

2 Find the centimeter mark that is closest to the other end.

To the nearest centimeter, the length of the piece of chalk

is _____ centimeters long.

Cut and use this centimeter ruler.

Online Content at **connectED.mcgraw-hill.com**

Try It

Find the length of the toy car to the nearest millimeter.

1 Place the ruler against one edge of the car. Line up the zero on the ruler with the end of the car.

2 Find the millimeter mark that is closest to the other end.

To the nearest millimeter, the toy car is _____ millimeters long.

Talk About It

1. Explain how you can tell the difference between the centimeter and millimeter marks when measuring an object with a metric ruler.

_____

_____

2. Is it easier to measure objects to the nearest centimeter or to the nearest millimeter? Explain.

_____

_____

3. **Mathematical PRACTICE 3** **Justify Conclusions** Should you measure the length across a penny to the nearest centimeter or millimeter? Explain your reasoning.

_____

_____

Practice It

Measure the length of each object to the nearest centimeter and millimeter.

4.

5.

_____

_____

Find the length of each object to the nearest centimeter and millimeter.

6. width of a book _____

7. length of a pencil _____

8. width of a calculator _____

9. length of a tape dispenser _____

Draw a line segment with each of the following lengths.

10. 6 centimeters

11. 27 millimeters

12. 5 centimeters

13. Compare the units of length you would use to measure the following: the length of a bicycle and the width of a dime. Explain your reasoning.

Ten speed

_____

_____

_____

_____

14. The length of a cell phone is 8 centimeters to the nearest centimeter and 81 millimeters to the nearest millimeter. Which measurement is more precise?

_____

Mathematical
15. PRACTICE ➌ Find the Error Jessie used a ruler to measure a colored pencil. Jessie said the pencil is 14.3 millimeters long. Find her mistake and correct it.

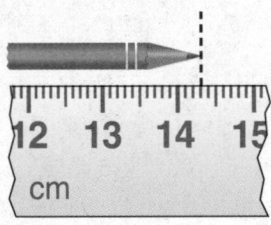

_____

_____

Write About It

16. Will I get a more precise measurement if I measure an object to the nearest centimeter or to the nearest millimeter? Explain your reasoning.

_____

_____

MY Homework

Homework Helper

Need help? connectED.mcgraw-hill.com

Find the length of the leaf to the nearest centimeter and millimeter.

1. Place the ruler against one edge of the object. Line up the zero on the ruler with the end of the object.

2. Find the centimeter and millimeter mark that is closest to the other end.

To the nearest centimeter, the leaf is 5 centimeters long. To the nearest millimeter, it is 48 millimeters long.

Practice

Measure the length of each object to the nearest centimeter and millimeter.

1.

2.

Find the length of each object to the nearest centimeter and millimeter.

3. length of a pen

4. length of a paper clip

_____

_____

Draw a line segment with each of the following lengths.

5. 7 centimeters

6. 105 millimeters

Problem Solving

7. **Mathematical PRACTICE 6** **Be Precise** The length of Manny's hamster is 114 millimeters to the nearest millimeter and 11 centimeters to the nearest centimeter. Which measurement is more precise?

_____

8. Lauren has a ruler that is marked in millimeters and a tape measure that is marked in centimeters. Which measuring tool will give Lauren a more precise measurement?

_____

9. Hanley measured the height of his glass to be 13 centimeters. Sally measured the same glass and found that it measured 132 millimeters. Who used a more precise measurement?

_____

Convert Metric Units of Length

Lesson 10

ESSENTIAL QUESTION
How can I use measurement conversions to solve real world problems?

The **metric system** is a decimal system of measurement. To convert metric units, multiply or divide by powers of 10.

 ## Math in My World

Example 1

One of the largest recorded pythons measured 7.3 meters long. What is the length of the python in centimeters?

Convert 7.3 meters to centimeters.

Since 1 meter = 100 centimeters, multiply 7.3 by 100.

So long!

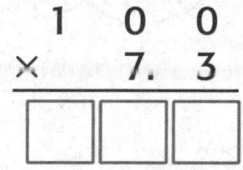

$$
\begin{array}{r}
1\ 0\ 0 \\
\times\ 7.\ 3 \\
\hline
\square\ \square\ \square
\end{array}
$$

To multiply by 10, 100, or 1,000, use basic facts and count the number of zeros in the factors.

So, 7.3 meters = _____ centimeters.

The python is _____ centimeters long.

Key Concept Metric Units of Length

1 **centimeter** (cm) = 10 **millimeters** (mm)

1 **meter** (m) = 100 cm or 1,000 mm

1 **kilometer** (km) = 1,000 m

1 millimeter thickness of a dime	**1 centimeter** width of pinky finger	**1 meter** height of a doorknob	**1 kilometer** 6 city blocks

Online Content at ⌁ **connectED.mcgraw-hill.com**

Example 2

Roshonda has 50 dominoes. Each domino is 4 centimeters long. She lines them up end to end as shown. How many meters long is the line of dominoes?

4 cm   4 cm   4 cm

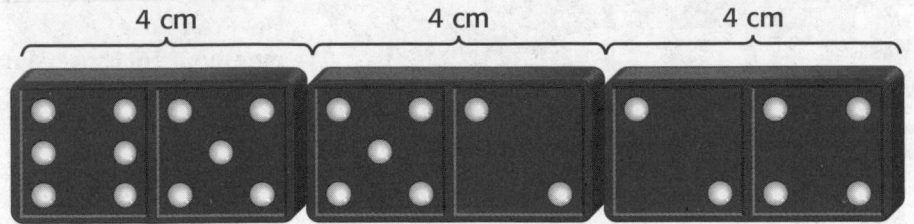

1 Find the length in centimeters.

50 × 4 centimeters = _____ centimeters

2 Convert _____ centimeters to meters.

Since 1 meter = _____ centimeters,

divide _____ by _____ .

_____ ÷ _____ = _____

So, _____ centimeters = _____ meters.

The line of 50 dominoes is _____ meters long.

Helpful Hint

To divide by 10, 100, or 1,000, cross out the same number of zeros in both the dividend and divisor.

Guided Practice

Complete.

1. 5 m = ■ cm

5 × 100 = _____

So, 5 meters equals _____ centimeters.

2. 9,000 m = ■ km

9,000 ÷ 1,000 = _____

So, 9,000 meters equals

_____ kilometers.

Talk MATH

How can you use mental math to convert 7.38 kilometers to meters?

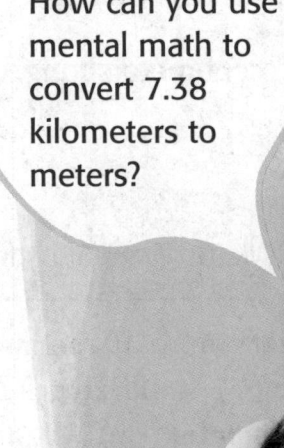

Independent Practice

Complete.

3. 700 cm = _____ m

4. 8,500 mm = _____ m

5. 15 km = _____ m

6. 73,000 m = _____ km

7. 2.71 m = _____ mm

8. 9.2 m = _____ cm

9. 17.5 mm = _____ cm

10. 0.509 km = _____ m

Complete. Use <, >, or = to make a true statement.

11. 30 cm ◯ 300 mm

12. 4.8 km ◯ 4,800 m

13. 25 mm ◯ 3 cm

14. 9 km ◯ 8,500 m

15. 1.5 m ◯ 145 cm

16. 17 m ◯ 116 cm

Problem Solving

17. Measure the distance across the sunflower to the nearest centimeter. How many centimeters shorter than 1 meter is the width of the sunflower?

_____

18. Mathematical PRACTICE 1 Check for Reasonableness Which is the most reasonable estimate for the depth of a lake: 6 millimeters, 6 centimeters, or 6 meters? Explain.

_____

_____

19. A spider is 6 millimeters long. What fractional part of 1 centimeter is 6 millimeters?

_____

_____

HOT Problems

20. Mathematical PRACTICE 3 Which One Doesn't Belong? Circle the measure that does not belong with the other three. Explain your reasoning.

_____

| 3,500 Km | 3.5 m | 350 cm | 3,500 mm |

21. **Building on the Essential Question** Compare and contrast converting customary units of length and converting metric units of length.

_____

_____

_____

Itsy, bitsy millimeter!

MY Homework

Homework Helper

Need help? connectED.mcgraw-hill.com

The average length of a great white shark is about 4 meters. What is the average length in centimeters?

Convert 4 meters to centimeters.

Since 1 meter = 100 centimeters, multiply 4 by 100.

4 × 100 = 400

So, 4 meters = 400 centimeters.

The average length of the great white shark is about 400 centimeters.

Practice

Complete.

1. 300 cm = _____ m

2. 500 mm = _____ cm

3. 1.7 km = _____ cm

4. 2 km = _____ m

5. 6 cm = _____ mm

6. 238 cm = _____ m

7. 2,400 mm = _____ m

8. 175 mm = _____ m

Problem Solving

9. When completed, a tunnel will be 1,500 meters long. What is this length in kilometers?

_____

10. Mathematical **PRACTICE** 2 **Use Number Sense** The depth of a swimming pool is 8.5 meters. What is half of the depth in millimeters?

_____

Vocabulary Check

Choose the correct word(s) that completes each sentence.

millimeter   centimeter   meter

kilometer   metric system

11. The _____ is an appropriate unit to measure the length of a ladybug.

12. The _____ is an appropriate unit to measure the distance between two cities.

13. The _____ is a decimal system of measurement.

Test Practice

14. Kaelyn is reading a book. The book's thickness is 31 millimeters. Which is the correct thickness in centimeters?

  Ⓐ 3.001 centimeters

  Ⓑ 3.01 centimeters

  Ⓒ 3.1 centimeters

  Ⓓ 3.11 centimeters

Check My Progress

Vocabulary Check

Choose the correct word(s) that completes each sentence.

capacity	centimeter	cup	fluid ounces	gallon
kilometer	pint	quart	meter	millimeter

1. The _____ is an appropriate unit to measure the height of an oak tree.

2. The _____ is an appropriate unit to measure the capacity of a gasoline tank.

Concept Check

Compare. Use >, <, or = to make a true statement.

3. 7 c ◯ 3¼ pt

4. 5 gal ◯ 18 qt

5. 45 cm ◯ 450 mm

6. 4.5 km ◯ 5,000 m

Complete.

7. 7 c = _____ fl oz

8. 17 gal = _____ qt

9. 22 qt = _____ gal _____ qt

10. 835 cm = _____ m

11. 88,000 m = _____ km

12. 49.3 mm = _____ cm

13. Make a line plot of the measurements in the table. Then find the fair share.

Board Lengths (ft)				
$\frac{1}{2}$	$\frac{1}{2}$	$\frac{1}{3}$	$\frac{1}{3}$	$\frac{1}{3}$

⟵————————————⟶

fair share: _____

Problem Solving

14. Which is the most reasonable estimate for the height of a two-story house: 15 centimeters, 15 meters, or 15 kilometers? Explain.

_____

15. Torri measured the capacity of the punch bowl. Her first measurement was 2 gallons. Her second measurement was 9 quarts. Compare the two measurements. Use >, <, or = to make a true statement.

_____

16. Phil has 7 quarts of hot chocolate to give to his classmates. How many of Phil's classmates can have one cup of hot chocolate?

_____

Test Practice

17. The depth of a lake is 1,400 meters. What is the depth in kilometers?

   Ⓐ 0.14 kilometers   Ⓒ 14 kilometers

   Ⓑ 1.4 kilometers   Ⓓ 140 kilometers

Hands On
Estimate and Measure Metric Mass

Lesson 11

ESSENTIAL QUESTION
How can I use measurement conversions to solve real-world problems?

The **mass** of an object is the amount of matter it has. A **gram** is a metric unit of measurement of mass.

Measure It

1 Estimate the mass of each object in grams. Record your results in the table.

Object	Mass (g)	
	Estimate	**Actual**
Scissors		
Pencil		
Stapler		
Calculator		

2 Measure the mass of each object.

Place the scissors on one side of a balance. Set gram weights on the other side until the sides are level. Record the actual mass. Repeat this step for the other objects.

Try It

A **kilogram** is also a metric unit of measurement of mass. One kilogram is equal to 1,000 grams. Use this to complete the table below.

Kilograms	Grams
1	1,000
2	
3	
4	
5	

Look for a pattern in the table.

How many grams are in 6 kilograms? _____

How many grams are in 9 kilograms? _____

Talk About It

1. Order the four objects you weighed in the first activity from greatest to least mass.

   _____

   Mathematical
2. **PRACTICE** 6 **Explain to a Friend** Use the mass of the objects you found to estimate the mass of two other objects in your classroom. Then find the mass of the objects. Were your estimates close?

   _____

   _____

3. Could a larger object have less mass than a smaller object? Explain.

   _____

   _____

4. Explain how you can use mental math in order to convert kilograms to grams.

   _____

Practice It

5. Identify three objects in your classroom that you can use the balance to find their masses. Estimate each object's mass. Then find the mass of each object and record the exact mass in the table.

Object	Mass (g)	
	Estimate	Actual

Compare. Use >, <, or = to make a true statement.

6. 1,500 grams $\bigcirc$ 1 kilogram

7. 3,000 grams $\bigcirc$ 3 kilograms

8. 4,000 grams $\bigcirc$ 3 kilograms

9. 3,700 grams $\bigcirc$ 4 kilograms

10. 5 kilograms $\bigcirc$ 6,000 grams

11. 3.5 kilograms $\bigcirc$ 3,000 grams

12. 2.5 kilograms $\bigcirc$ 2,500 grams

13. 3.25 kilograms $\bigcirc$ 3,300 grams

14. Randy and Salma measured the mass of the same chinchilla. Randy measured the chinchilla as 1 kilogram. Salma measured the chinchilla as 945 grams. Circle the more precise measurement.

945 grams   1 kilogram

15. Edita measured the mass of her books. She measured the mass as 2 kilograms. Her second measurement was 2,050 grams. Use >, <, or = to make a true statement.

2 kilograms ◯ 2,050 grams

16. Mathematical PRACTICE 6 Be Precise If you are measuring the mass of a container of salt, would grams or kilograms give you a more precise measurement? Explain.

_____

_____

_____

17. Mathematical PRACTICE 3 Draw a Conclusion Compare and contrast grams and kilograms.

_____

_____

_____

Write About It

18. How can I convert grams to kilograms without measuring?

_____

My Work!

Name _____

MY Homework

Homework Helper

Need help? ☞ connectED.mcgraw-hill.com

One kilogram is equal to 1,000 grams. Use this information to complete the table. How many grams are in 6 kilograms?

For every increase of one kilogram, increase the number of grams by 1,000.

Kilograms	Grams
1	1,000
2	2,000
3	3,000
4	4,000
5	5,000
6	6,000

+ 1,000
+ 1,000
+ 1,000
+ 1,000
+ 1,000

So, 6 kilograms are equal to 6,000 grams.

Practice

Compare. Use >, <, or = to make a true statement.

1. 2,300 grams ◯ 2 kilograms

2. 4,840 grams ◯ 5 kilograms

3. 4 kilograms ◯ 4,150 grams

4. 1.75 kilograms ◯ 1,750 grams

Vocabulary Check

5. Fill in the blank with the correct word to complete the sentence below.

The _____ of an object is the amount of matter it has.

Problem Solving

6. Jason and Scott measured the masses of their cell phones. Jason measured his cell phone using kilograms. Scott measured his cell phone using grams. Which measure would be more appropriate to measure a cell phone?

_____

7. **Mathematical** **PRACTICE** **6** **Be Precise** Gavin has a ten-year old cat, Shadow. Is Shadow's mass more likely to be 6 kilograms or 6 grams? Explain.

_____

_____

8. Norton measured the mass of his luggage. His luggage mass was 21,530 grams. The airline will only allow luggage that has a mass under 23 kilograms. Will Norton be allowed to fly with his luggage? Explain.

_____

9. Chasity measured the mass of her new puppy. Her first measurement was 2,350 grams. Her second measurement was 2.3 kilograms. Circle the measurement that is more precise.

    2,350 grams   2.3 kilograms

10. Billy measured the mass of his iguana. His first measurement was 4,100 grams. His second measurement was 4 kilograms. Compare the two measurements. Use >, <, or = to make a true statement.

_____

My Work!

Heavy dude!

Convert Metric Units of Mass

Lesson 12

ESSENTIAL QUESTION
How can I use measurement conversions to solve real-world problems?

Mass is a measure of the amount of matter an object has.

 ## Math in My World

136 kilograms
OH DEAR!

Example 1

A white-tailed deer has a mass of 136 kilograms. What is the mass of the deer in grams?

Convert 136 kilograms to grams.

Since 1 kilogram = 1,000 grams, multiply 136 by 1,000.

So, 136 kilograms = _____ grams.

$$\begin{array}{r} 1,000 \\ \times\ \ 136 \\ \hline 136,000 \end{array}$$

The mass of the white-tailed deer is _____ grams.

Check Use division to check your answer.

_____ ÷ 1,000 = 136

Key Concept Metric Units of Mass

1 **gram** (g) = 1,000 **milligrams** (mg)   1 **kilogram** (kg) = 1,000 g

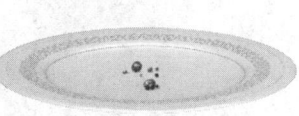

1 milligram
a bread crumb

1 gram
a paper clip

1 kilogram
a loaf of bread

Online Content at connectED.mcgraw-hill.com

Example 2

Convert 1,500 grams to kilograms.

Since you are converting a smaller unit to a larger unit, divide.

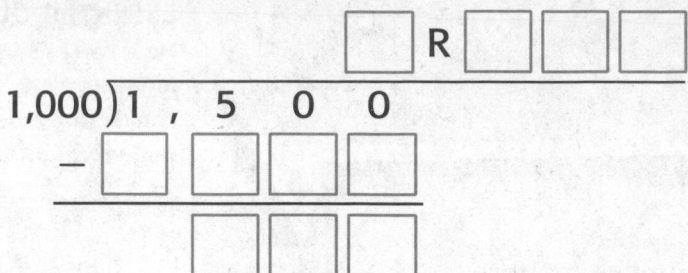

The remainder _____ means there are _____ grams left over.

The decimal part of a kilogram is _____.

So, 1,500 grams = _____ kilogram _____ grams or _____ kilograms.

Guided Practice

Complete.

1. 5,000 mg = ■ g

5,000 ÷ 1,000 = _____

So, 5,000 milligrams equals

_____ grams.

2. 5 kg = ■ g

5 × 1,000 = _____

So, 5 kilograms equals

_____ grams.

3. 4,000 g = ■ kg

4,000 ÷ 1,000 = _____

So, 4,000 grams equals

_____ kilograms.

4. 9 g = ■ mg

9 × 1,000 = _____

So, 9 grams equals

_____ milligrams.

Talk MATH

Which is a more reasonable estimate for the mass of a baseball: 140 milligrams, 140 grams, or 140 kilograms? Explain.

Independent Practice

Complete.

5. 2,000 mg = _____ g

6. 80 g = _____ mg

7. 0.75 kg = _____ mg

8. 6 kg = _____ g

9. 3,100 g = _____ kg

10. 0.05 kg = _____ mg

11. 4.07 g = _____ mg

12. 9 kg = _____ g

Compare. Use <, >, or = to make a true statement.

13. 2,300 mg ◯ 2 g

14. 3 kg ◯ 3,000 g

15. 4.5 kg ◯ 4,050 g

16. 4,120 mg ◯ 4.12 g

17. 75 g ◯ 800 mg

18. 814 g ◯ 8.14 kg

Problem Solving

Use the table shown for Exercises 19–21.

Macaws	
Species	Mass (grams)
Blue and Gold	800
Green-winged	900
Red-footed	525
Yellow-collared	250

19. How many yellow-collared macaws would have a combined mass of 1 kilogram?

_____

Mathematical
20. PRACTICE 6 **Explain to a Friend** Is the combined mass of two red-footed macaws and three blue and gold macaws closer to 3 kilograms or 4 kilograms? Explain.

_____

_____

21. Which macaw has a mass closest to 1 kilogram?

_____

Pretty bird!

 **HOT Problems**

Mathematical
22. PRACTICE 2 **Use Number Sense** One pound is approximately equal to 0.5 kilograms. About how many kilograms is 3 pounds?

_____

23. **Building on the Essential Question** How is converting metric units of mass different from converting customary units of weight?

_____

_____

MY Homework

Homework Helper

Need help? connectED.mcgraw-hill.com

**Mr. Benavides bakes muffins that have a mass of about
50,000 milligrams. What is the mass in grams?**

Convert 50,000 milligrams to grams.

Since 1,000 milligrams = 1 gram, divide 50,000 by 1,000.

So, 50,000 milligrams = 50 grams.

The muffins have a mass of about 50 grams.

Practice

Complete.

1. 7,000 mg = _____ g

2. 4.7 kg = _____ g

3. 18,500 g = _____ kg

4. 8.3 kg = _____ g

5. 22 g = _____ mg

6. 135,000 mg = _____ kg

Problem Solving

7. One highlighter has a mass of 11 grams. Another highlighter has a mass of 10,800 milligrams. Which highlighter has the greater mass?

_____

Mathematical
8. PRACTICE  **Be Precise** One computer has a mass of 0.8 kilogram and another has a mass of 800 grams. Compare the masses of the computers. Use >, <, or = to make a true statement.

_____

Vocabulary Check

Fill in the correct circle that corresponds to the best answer.

9. Which of the following is not a common unit of measurement for the metric system?

   Ⓐ milligram   Ⓒ gram

   Ⓑ kilogram   Ⓓ ounce

10. Which operation is necessary to convert a larger unit to a smaller unit?

   Ⓕ addition   Ⓗ multiplication

   Ⓖ subtraction   Ⓘ division

Test Practice

11. For a science experiment, Noel measured a piece of metal that has a mass of 3,500 grams. What is the mass of the metal in kilograms?

   Ⓐ 0.35 kilogram   Ⓒ 35 kilograms

   Ⓑ 3.5 kilograms   Ⓓ 350 kilograms

Convert Metric Units of Capacity

Lesson 13

ESSENTIAL QUESTION
How can I use measurement conversions to solve real-world problems?

In the metric system, the common units of capacity are liter and milliliter.

 **Math in My World**

Example 1

A dripping faucet wastes about 90 liters of water every week. How many milliliters of water is this?

Convert 90 liters to milliliters.

Since 1 liter = 1,000 milliliters, multiply 90 by 1,000.

$$
\begin{array}{r}
1{,}000 \\
\times\ \ \ \ 90 \\
\hline
90{,}000
\end{array}
$$

So, 90 liters = _____ milliliters.

The dripping faucet wastes _____ milliliters of water.

Key Concept Metric Units of Capacity

1 **liter** (L) = 1,000 **milliliters** (mL)

1 milliliter
amount of liquid in
an eyedropper

1 liter
a medium-sized
sports drink

Online Content at 🖰 **connectED.mcgraw-hill.com**

Example 2

A container of orange juice holds 580 milliliters. How many liters is 580 milliliters?

Since 1 liter = _____ milliliters, divide 580 by _____ .

580 ÷ _____ = _____   ← Move the decimal point 3 places to the left.

So, 580 milliliters = _____ liter.

The container holds _____ liter of orange juice.

Guided Practice

Complete.

1. 6 L = ■ mL

6 × 1,000 = _____

So, 6 liters equals _____ milliliters.

2. 4 L = ■ mL

4 × 1,000 = _____

So, 4 liters equals _____ milliliters.

3. 7,000 mL = ■ L

7,000 ÷ 1,000 = _____

So, 7,000 milliliters equals _____ liters.

4. 42 mL = ■ L

42 ÷ 1,000 = _____

So, 42 milliliters equals _____ liter.

Which unit would you use to measure the capacity of a glass of milk: milliliter or liter? Explain.

Independent Practice

Complete.

5. 70 L = _____ mL

6. 10 mL = _____ L

7. 1.2 L = _____ mL

8. 3,500 mL = _____ L

9. 4 L = _____ mL

10. 230 mL = _____ L

11. 6.21 L = _____ mL

12. 5,000 mL = _____ L

Compare. Use <, >, or = to make a true statement.

13. 2 L ◯ 1,000 mL

14. 390 mL ◯ 0.39 L

15. 82 L ◯ 825 mL

16. 834 mL ◯ 8.34 L

17. 0.34 L ◯ 430 mL

18. 87 mL ◯ 0.087 L

Problem Solving

19. The Nail Shop purchases nail polish in 13-milliliter bottles. Find the total capacity, in liters, of 1,000 bottles.

_____

20. Paulene measures the water in a container to be 2,732 milliliters. Justin measures the water in the same container to be 3 liters. Circle the greater measurement.

2,732 milliliters   3 liters

21. Mathematical **PRACTICE** 1 **Check for Reasonableness** To prepare for his camping trip, Emanuel filled his canteen with water. Is 15,000 milliliters or 1,500 milliliters a more reasonable estimate for the amount of water in the canteen? Explain.

_____

_____

_____

I'm all filled up!

HOT Problems

22. Mathematical **PRACTICE** 2 **Reason** Name three items that have a capacity greater than 10 liters.

_____

23. ❓ **Building on the Essential Question** Why is it important to be able to convert metric units of capacity?

_____

_____

MY Homework

Homework Helper

Need help? connectED.mcgraw-hill.com

A cough syrup bottle contains 120 milliliters of cough syrup. How many liters is 120 milliliters?

Since 1 liter = 1,000 milliliters, divide 120 by 1,000.

120 ÷ 1,000 = 0.12 ◀— Move the decimal point 3 places to the left.

So, 120 milliliters = 0.12 liter.

The bottle holds 0.12 liter of cough syrup.

Practice

Complete.

1. 6 L = _____ mL

2. 13 L = _____ mL

3. 54,000 mL = _____ L

4. 23,500 mL = _____ L

5. 11,000 mL = _____ L

6. 0.201 L = _____ mL

Problem Solving

7. Yesterday, Audrey drank the liquids shown in the table. How many liters of liquids did she drink in all?

Liquid	Amount
Juice	210 mL
Milk	480 mL
Water	1.2 L

_____

8. One serving of punch is 250 milliliters. Will ten servings fit in a 2-liter bowl? Explain.

_____

_____

9. **Mathematical PRACTICE** **Make Sense of Problems** Nara received a measles immunization at Dr. Arroyo's office. The vaccine was measured in cubic centimeters. A cubic centimeter has the same capacity as a milliliter. If the immunization was 3.5 cubic centimeters, how many milliliters was it?

_____

Vocabulary Check

Fill in the blank with the correct word(s) that completes each sentence.

10. The _____ is an appropriate unit to measure the capacity of a hand sanitizer bottle.

11. The _____ is an appropriate unit to measure the capacity of the water in a fountain.

Test Practice

12. A soup bowl can hold about 400 milliliters of soup. A restaurant has 8 liters of vegetable soup. How many bowls of soup can they serve?

Ⓐ 500 bowls   Ⓒ 50 bowls

Ⓑ 200 bowls   Ⓓ 20 bowls

Review

Vocabulary Check

Fill in the circle next to the best answer.

1. The **capacity** of a container is which of the following?

   Ⓐ the elapsed time

   Ⓑ the customary unit

   Ⓒ the metric unit

   Ⓓ the amount that it can hold

2. Customary units of **length** are measured in which of the following?

   Ⓕ meters and centimeters only

   Ⓖ inches, feet, yards, and miles

   Ⓗ minutes and hours

   Ⓘ days and weeks

3. The **metric system** is based on which of the following?

   Ⓐ fractions

   Ⓑ decimals

   Ⓒ inches

   Ⓓ gallons

4. When you **convert** from feet to inches, you are doing which of the following?

   Ⓕ changing the measurement unit

   Ⓖ determining capacity

   Ⓗ determining mass

   Ⓘ determining volume

5. When finding the **mass** of an object, you determine which of the following?

   Ⓐ the quantity of matter in the object

   Ⓑ its weight

   Ⓒ its height

   Ⓓ its length

Concept Check

Complete.

6. 84 in. = _____ ft

7. 9 yd = _____ in.

8. 7,920 yd = _____ mi

9. 64,000 lb = _____ T

10. $7\frac{1}{2}$ lb = _____ oz

11. 62 oz = _____ lb _____ oz

12. 7 pt = _____ c

13. 12 c = _____ qt

14. 72 pt = _____ gal

15. 120 mm = _____ cm

16. Make a line plot of the measurements in the table. Then find the fair share.

Amount of Sports Drink (gal)							
$\frac{1}{4}$	$\frac{1}{2}$	$\frac{1}{3}$	$\frac{1}{2}$	$\frac{1}{4}$	$\frac{1}{3}$	$\frac{1}{3}$	$\frac{1}{2}$

⟵——————————————⟶

fair share: _____

Problem Solving

17. Breanna has quarters, dimes, and nickels in her purse. She has
3 fewer nickels than dimes, but she has 2 more nickels than quarters.
If Breanna has 2 quarters, how much money does she have?

_____

18. A detergent bottle holds 700 milliliters. Find the capacity in liters.

_____

19. When Quentin flew from New York City to Atlanta, the pilot
announced that they were flying at 33,000 feet. How many miles is
this? Write as a mixed number.

_____

20. Deka measured the mass of 100 sheets of paper as 1,500 grams.
How many kilograms is this?

_____

Test Practice

21. Maisha is using special paint for her artwork. The art supply store
charges $1.50 per cup of paint. Maisha needs 2 pints of blue paint,
3 cups of green paint, $1\frac{1}{2}$ quarts of orange paint, and $\frac{1}{2}$ cup of
yellow paint. How much will she pay?

    Ⓐ $10.50   Ⓒ $14.25

    Ⓑ $11.25   Ⓓ $20.25

Reflect

Use what you learned about measurement to complete
the graphic organizer below.

ESSENTIAL QUESTION

How can I use measurement conversions to solve real-world
problems?

0 1 2 3 inches **Customary System**	0 1 2 3 4 5 mm **Metric System**
Vocabulary	**Vocabulary**
Conversions	**Conversions**

Now reflect on the ESSENTIAL QUESTION Write your answer below.

_____

_____

ESSENTIAL QUESTION

How does geometry help me solve problems in everyday life?

Let's Travel!

Watch a video!

Watch

MY Standards

Geometry

5.G.3 Understand that attributes belonging to a category of two-dimensional figures also belong to all subcategories of that category.

5.G.4 Classify two-dimensional figures in a hierarchy based on properties.

Measurement and Data *This chapter also addresses these standards:*

5.MD.3 Recognize volume as an attribute of solid figures and understand concepts of volume measurement.

5.MD.3a A cube with side length 1 unit, called a "unit cube," is said to have "one cubic unit" of volume, and can be used to measure volume.

5.MD.3b A solid figure which can be packed without gaps or overlaps using n unit cubes is said to have a volume of n cubic units.

5.MD.4 Measure volumes by counting unit cubes, using cubic cm, cubic in, cubic ft, and improvised units.

5.MD.5 Relate volume to the operations of multiplication and addition and solve real world and mathematical problems involving volume.

5.MD.5a Find the volume of a right rectangular prism with whole-number side lengths by packing it with unit cubes, and show that the volume is the same as would be found by multiplying the edge lengths, equivalently by multiplying the height by the area of the base. Represent threefold whole-number products as volumes, e.g., to represent the associative property of multiplication.

5.MD.5b Apply the formulas $V = l \times w \times h$ and $V = b \times h$ for rectangular prisms to find volumes of right rectangular prisms with whole-number edge lengths in the context of solving real world and mathematical problems.

5.MD.5c Recognize volume as additive. Find volumes of solid figures composed of two non-overlapping right rectangular prisms by adding the volumes of the non-overlapping parts, applying this technique to solve real world problems.

Standards for Mathematical PRACTICE

I'll be able to get this - no problem!

1. Make sense of problems and persevere in solving them.
2. Reason abstractly and quantitatively.
3. Construct viable arguments and critique the reasoning of others.
4. Model with mathematics.
5. Use appropriate tools strategically.
6. Attend to precision.
7. Look for and make use of structure.
8. Look for and express regularity in repeated reasoning.

 = focused on in this chapter

Name _____

Am I Ready?

Check ✓ ← Go online to take the Readiness Quiz

Name the number of sides and the number of angles in each figure.

1.

_____ sides and _____ angles

2.

_____ sides and _____ angles

3.

_____ sides and _____ angles

4.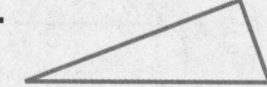

_____ sides and _____ angles

Use the figure below for Exercises 5 and 6.

```
A _____ B
 |        |
 |        |
D _____ C
```

5. Which side appears to have the same length as side *AD*? _____

6. At which point do sides *AB* and *BC* meet? _____

7. Kai is drawing a triangle with all three sides that are equal. Draw a sketch of this triangle.

Shade the boxes to show the problems you answered correctly.

How Did I Do? → | 1 | 2 | 3 | 4 | 5 | 6 | 7 |

MY Math Words

Review Vocabulary

acute angle	angles	lines	obtuse angle
parallel	perpendicular	right angle	

Making Connections

Use the review words to classify geometric shapes.

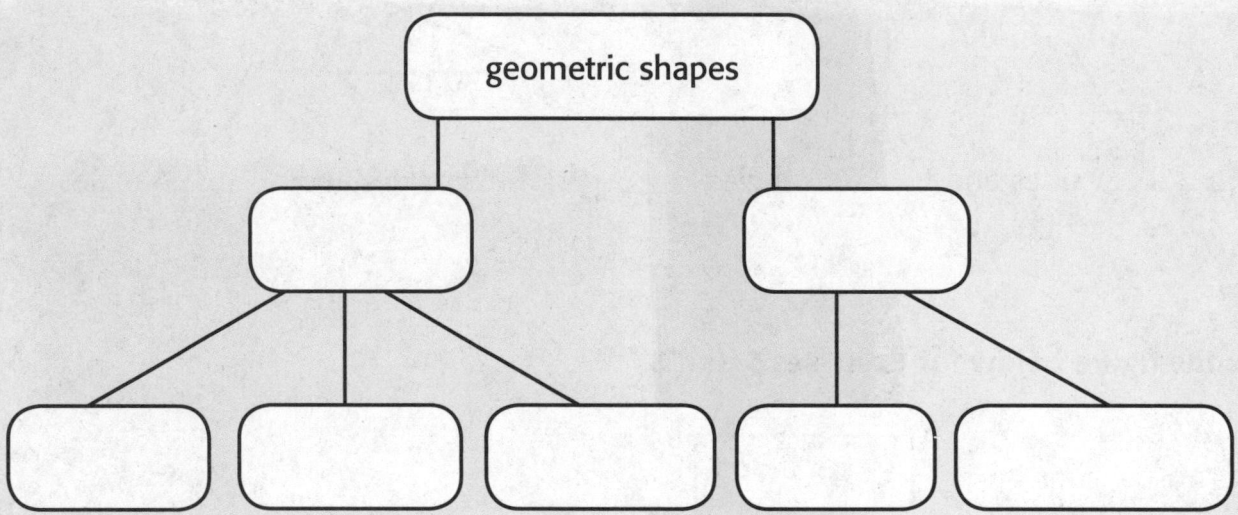

Draw an example of two of the words used above.

MY Vocabulary Cards

Lesson 12-3

acute triangle

Lesson 12-3

attribute

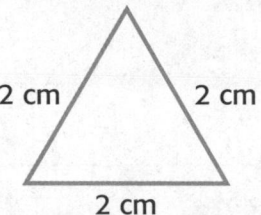

2 cm   2 cm

2 cm

Lesson 12-7

bases

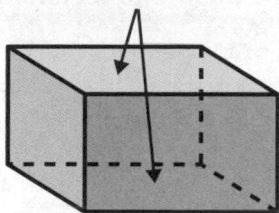

Lesson 12-10

composite figures

Lesson 12-1

congruent angles

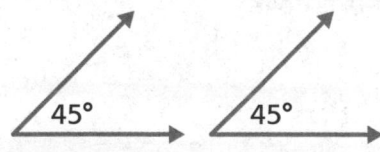

45°   45°

Lesson 12-6

congruent figures

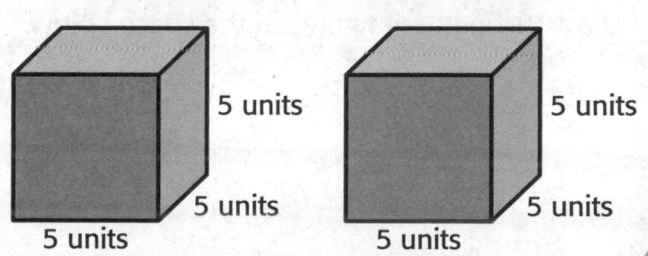

5 units   5 units

5 units   5 units

5 units   5 units

Lesson 12-1

congruent sides

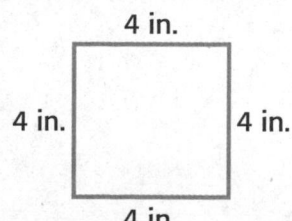

4 in.

4 in.   4 in.

4 in.

Lesson 12-6

cube

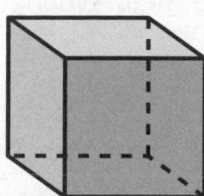

Ideas for Use

- Design a crossword puzzle. Use the definition for each word as the clues.

- Group 2 or 3 common words. Add a word that is unrelated to the group. Then work with a friend to name the unrelated word.

A characteristic of a figure.

Use *attribute* to describe the sides or angles of a rectangle.

_____

_____

A triangle with 3 acute angles.

Explain how to determine if a triangle is an acute triangle.

_____

_____

A figure that is made of two or more three-dimensional figures.

Composite comes from *compose,* "to put together." How does this help you understand a composite figure?

_____

_____

Two parallel congruent faces in a prism.

Describe the *base* of a rectangular prism.

_____

_____

_____

Two figures that have the same size and shape.

Draw 2 congruent figures in the space below.

Angles of a figure that are equal in measure.

Draw an example of a figure with congruent angles. Then draw a non-example. Cross out the non-example.

A three-dimensional figure with six faces that are congruent squares.

What does *cube* mean when it is used as a verb?

_____

_____

Sides of a figure that are equal in length.

Draw two figures that each have at least two congruent sides.

MY Vocabulary Cards

Lesson 12-8

cubic unit

4 cubic units

Lesson 12-7

edge

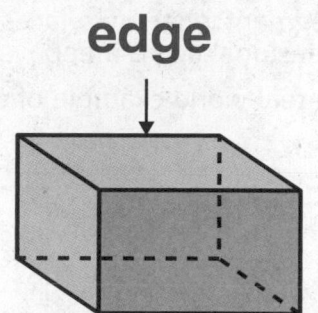

Lesson 12-3

equilateral triangle

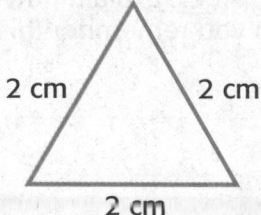

2 cm   2 cm

2 cm

Lesson 12-6

face

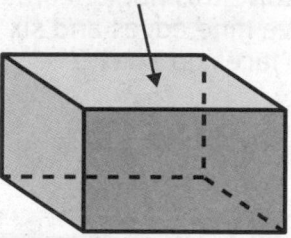

Lesson 12-1

hexagon

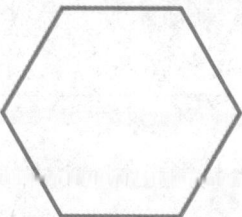

Lesson 12-3

isosceles triangle

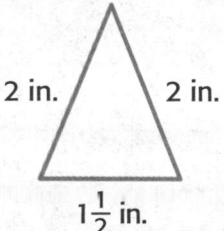

2 in.   2 in.

$1\frac{1}{2}$ in.

Lesson 12-6

net

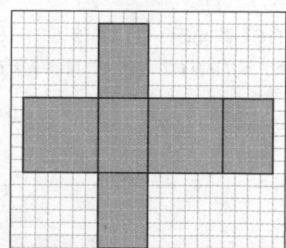

Lesson 12-3

obtuse triangle

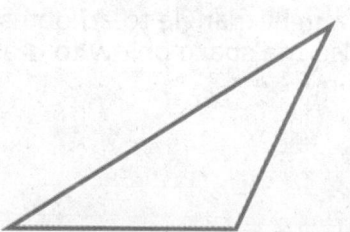

Ideas for Use

- Write a tally mark on each card every time you read the word in this chapter or use it in your writing. Challenge yourself to use at least 10 tally marks for each word card.

- Draw or write additional examples for each card. Be sure your examples are different from what is on the front of the card.

The line segment where two faces of a three-dimensional figure meet.

Describe a real-world example of an edge.

_____

_____

The unit of measure for volume.

Name three other units of measurement in math and what they measure.

_____

_____

A flat surface.

Read and solve this riddle: I am a triangular prism. I have nine edges and six vertices. How many faces do I have?

_____

A triangle with three congruent sides.

The prefix *equi-* means "equal." *Lat* is a Latin root meaning "side." Explain how these word parts can help you remember this definition.

_____

_____

A triangle with at least two congruent sides.

Draw an example of an isosceles triangle.

A polygon with six sides and six angles.

How are hexagons a subcatergory of polygons?

_____

_____

A triangle with 1 obtuse angle and 2 acute angles.

Compare a right triangle to an obtuse triangle. Use the space below to draw your comparison.

A two-dimensional pattern of a three-dimensional figure.

What does net mean in this sentence? *The soccer player's hand became tangled in the goal's net.*

_____

_____

MY Vocabulary Cards

Lesson 12-1

octagon

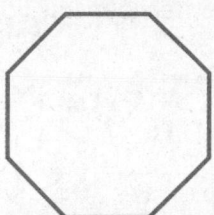

Lesson 12-5

parallelogram

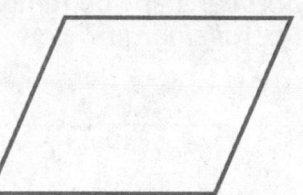

Lesson 12-1

pentagon

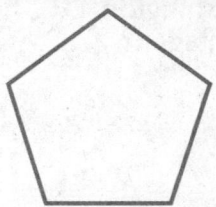

Lesson 12-1

polygon

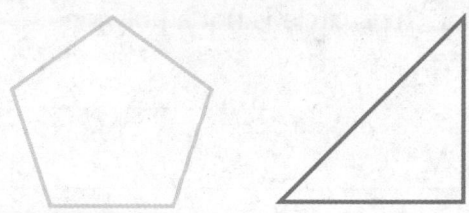

Lesson 12-7

prism

 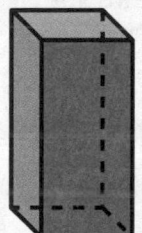

Lesson 12-5

rectangle

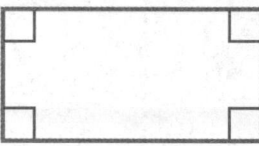

Lesson 12-6

rectangular prism

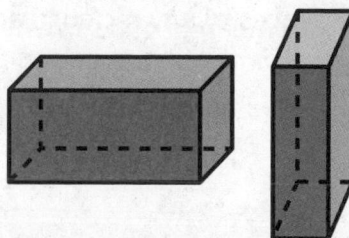

Lesson 12-1

regular polygon

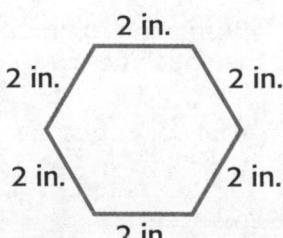

2 in.

2 in.   2 in.

2 in.   2 in.

2 in.

Ideas for Use

- Develop categories for the words. Sort them by category. Ask another student to guess each category.

- Draw or write examples for each card. Be sure your examples are different from what is shown on each card.

A quadrilateral in which each pair of opposite sides is parallel and congruent.

How does *parallel* help you remember the meaning of *parallelogram*?

_____

_____

A polygon with eight sides.

Okto is a Greek root meaning "eight." How can this help you remember this vocabulary word?

_____

_____

A closed figure made up of line segments that do not cross each other.

Explain why a circle is not a polygon.

_____

_____

A polygon with five sides.

How can the Pentagon, a government building in Washington, D.C., help you remember pentagon?

_____

_____

A quadrilateral with four right angles; opposite sides are equal and parallel.

Compare a rectangle with a square.

_____

_____

A three-dimensional figure with two parallel, congruent faces, called bases. At least three faces are rectangles.

Look at the orange prism on the front of the card. What shape are its bases?

_____

_____

A polygon in which all sides and angles are congruent.

What is one way to determine if a polygon is regular?

_____

_____

A prism that has rectangular bases.

Describe the faces of any rectangular prism.

_____

_____

MY Vocabulary Cards

 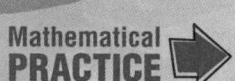

Lesson 12-5

rhombus

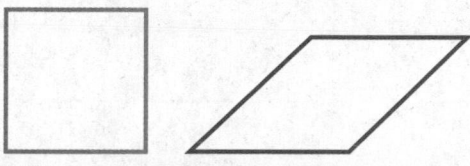

Lesson 12-3

right triangle

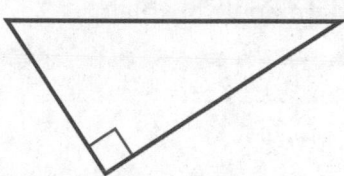

Lesson 12-3

scalene triangle

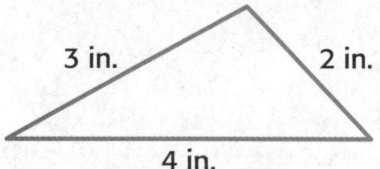

3 in.   2 in.

4 in.

Lesson 12-5

square

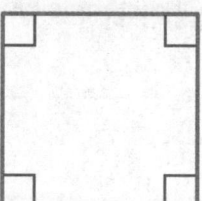

Lesson 12-6

three-dimensional figure

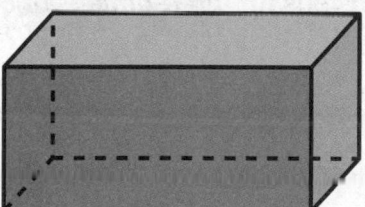

Lesson 12-5

trapezoid

Lesson 12-7

triangular prism

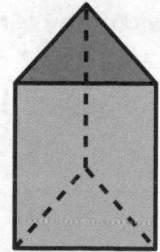

Lesson 12-8

unit cube

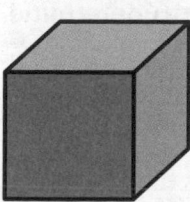

Ideas for Use

- Practice your penmanship! Write each word in cursive.

- Sort cards so that only polygons are displayed. Explain your sorting to a partner.

A triangle with 1 right angle and 2 acute angles.

Is it possible for a right triangle to have more than one right angle? Explain.

_____

_____

A parallelogram with four congruent sides.

Explain whether a rectangle is a rhombus.

_____

_____

A parallelogram with four congruent sides and four right angles.

Is a square also a rectangle? Explain.

_____

_____

A triangles with no congruent sides.

Draw a scalene triangle below.

A quadrilateral with exactly one pair of opposite sides parallel.

Draw a picture of two quadrilaterals—one that is a trapezoid and one that is not.

A figure that has length, width, and height.

Write a tip to help you remember the number of dimensions for *three-dimensional figures*.

_____

_____

A cube with a side length of one unit.

Draw a rectangular prism below that has a volume of 8 unit cubes.

A prism that has triangular bases.

Describe the bases of any triangular prism.

_____

_____

MY Vocabulary Cards

Mathematical PRACTICE

Lesson 12-7

vertex

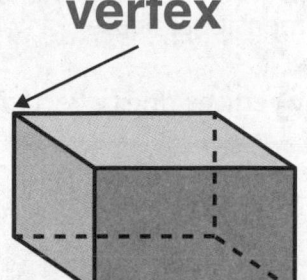

Lesson 12-8

volume

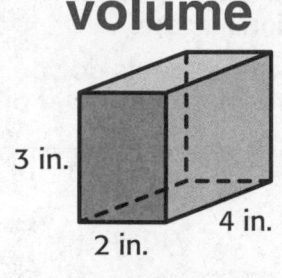

3 in.

2 in.

4 in.

$V = \ell wh$

Ideas for Use

- Write key concepts from some lessons on the front of blank cards. Write a few examples on the back of each card to help you study.

- Use a blank card to write this chapter's essential question. Use the back of the card to write or draw examples that help you answer the question.

The amount of space inside a three-dimensional figure.

How many measurements do you need to find the volume of a rectangular prism? What are they?

_____

The point where three or more faces meet on a three-dimensional figure.

How many vertices does a rectangular prism have?

_____

MY Foldable

FOLDABLES® Follow the steps on the back to make your Foldable.

Octagon

Not Regular

Regular

Hexagon

Not Regular

Regular

Pentagon

Not Regular

Regular

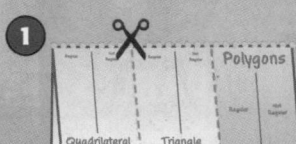

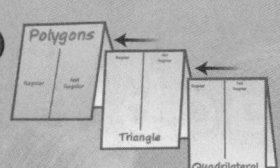

Polygons

Not Regular

Regular

Not Regular

Regular

Triangle

Not Regular

Regular

Quadrilateral

Polygons

Lesson 1

ESSENTIAL QUESTION
How does geometry help
me solve problems in
everyday life?

A **polygon** is a closed figure made up of line
segments that do not cross each other.

Polygons

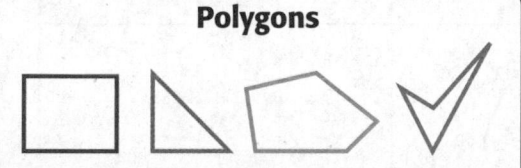

Not Polygons

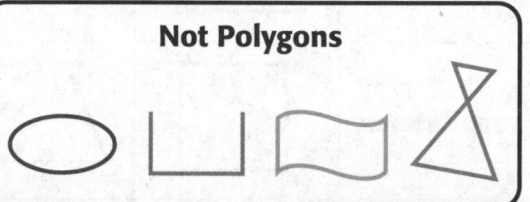

 ## Math in My World

Describe
my sides!

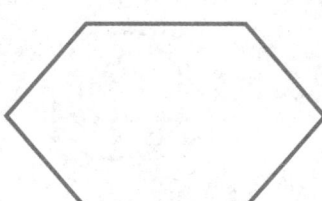

Example 1

**The building shown is the Pentagon in Washington,
D.C. Describe the sides of the figure formed by the
red outline. Does the red outline form a polygon?**

The figure has _____ sides.

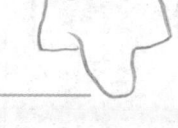

Do the sides ever cross each other? _____
The figure is a polygon.

A **regular polygon** is a polygon with congruent sides and congruent
angles. **Congruent sides** are equal in length. **Congruent angles** have
the same degree measure.

Example 2

**Determine if the polygon appears to be
regular or *not regular*.**

The top and bottom sides appear _____ than
the other sides.

Are all six sides of the polygon congruent? _____

It is _____ regular.

Polygons are a subcategory of two-dimensional figures. A *subcategory* is a subdivision that has common characteristics within a larger category.

Example 3

Kaleb

Complete the table below.

Polygon	Regular	Not Regular	Number of Sides	Draw another polygon that is not regular.
Triangle				
Quadrilateral				
Pentagon				
Hexagon				
Octagon				

Guided Practice

1. Name the polygon. Determine if it appears to be *regular* or *not regular*.

The polygon has ___8___ sides.

The sides appear to be ___line___ .

It is a ___octagoon___ .

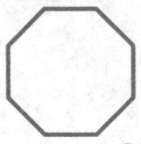

Talk MATH

Is a circle a polygon? Explain.

Independent Practice

Mathematical
PRACTICE **Identify Structure** Name each polygon. Determine if it appears to be *regular* or *not regular*.

2.

_____

3.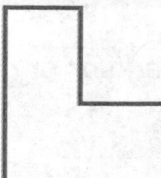

_____

4.

_____

5.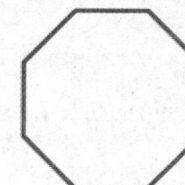

_____

Draw each polygon.

6. triangle; not regular

7. pentagon; not regular

8. quadrilateral; not regular

9. triangle; regular

Problem Solving

10. What polygons make up the design?

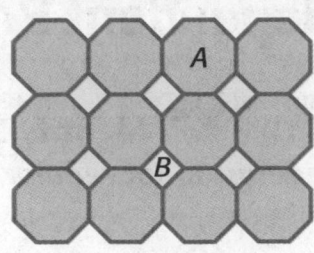

_____

11. Describe polygon B as *regular* or *not regular*.

_____

For Exercises 12 and 13, use the map shown at the right.

12. Circle the polygon that is a quadrilateral.

13. Describe polygon C as *regular* or *not regular*.

_____

HOT Problems

14. **Mathematical PRACTICE** **1** **Make Sense of Problems** Explain why every square is a regular polygon.

_____

_____

15. **Building on the Essential Question** How can polygons be considered a subcategory of two-dimensional figures?

_____

_____

_____

MY Homework

Homework Helper

Need help?  connectED.mcgraw-hill.com

Name the polygon used to form the greeting card shown. Does the red outline appear to be a regular polygon?

The polygon has four sides.

The top and bottom sides appear to be slightly longer than the other sides.

It is a quadrilateral.

It is not regular.

Practice

Name each polygon. Determine if it appears to be *regular* or *not regular.*

1.

2.

_____

Vocabulary Check

Fill in each blank with the correct word(s) to complete each sentence.

3. A polygon is a ___IDK___ figure made up of line segments that do not cross each other.

4. A regular polygon is a polygon with ___IDK___ sides and _____ angles.

Problem Solving

For Exercises 5–7, use the tangram pieces shown at the right.

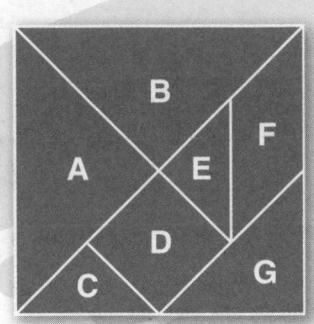

5. Which of the polygon(s) appear to be regular?

_____

6. What polygons are represented in the tangrams?

_____

7. Congruent figures have the same size and shape. Which polygons appear to be congruent?

_____

8. Name the polygon used to form the front of the tent shown. Determine if it appears to be *regular* or *not regular.*

_____

Mathematical
9. PRACTICE 1 **Make Sense of Problems**
Explain why the figure is not a polygon.

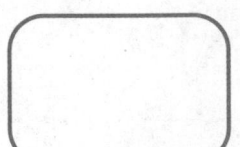

_____

_____

Test Practice

10. Which of the following figures is a polygon?

Ⓐ    Ⓑ 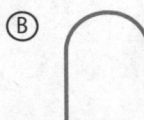   Ⓒ    Ⓓ

Hands On
Sides and Angles of Triangles

Lesson 2

ESSENTIAL QUESTION
How does geometry help me solve problems in everyday life?

A triangle is a polygon with three sides and three angles.

Measure It Tools

Measure the sides of each pair of triangles below to the nearest tenth of a centimeter. Then record the measures.

Pair A

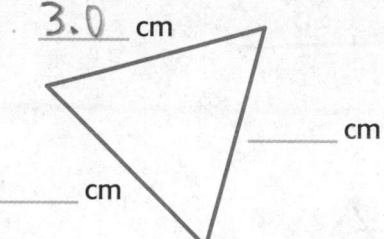

_3.0_ cm

____ cm

____ cm

____ cm

____ cm

____ cm

Pair B

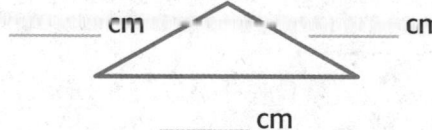

____ cm

____ cm

____ cm

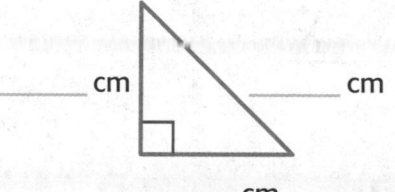

____ cm

____ cm

____ cm

Pair C

____ cm

____ cm

____ cm

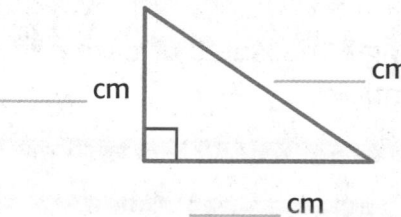

____ cm

____ cm

____ cm

Talk About It

1. Compare the side lengths of each pair of triangles above. What do you notice?

_____

_____

Cut and use this centimeter ruler. →

cm 0 1 2 3 4 5

Try It

Measure the angles of each pair of triangles below to the nearest degree. Then record the measures.

Pair A

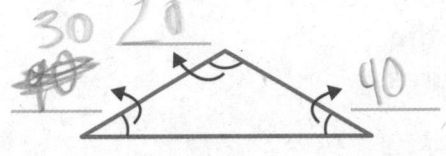

19

90   60

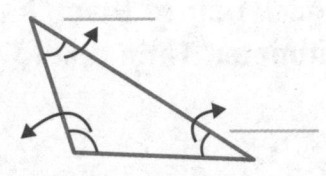

90

90   70

Pair B

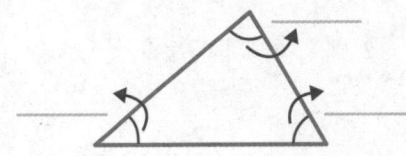

30 20

40

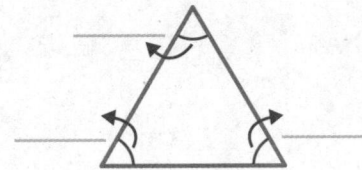

Pair C

Talk About It

2. Compare the angle measures of each pair of triangles above. What do you notice?

_____

_____

3. **Mathematical PRACTICE** **1** **Make Sense of Problems** Explain how a triangle is a special kind of polygon.

_____

_____

Practice It

Measure the sides of each triangle to the nearest tenth of a centimeter. Then describe the number of congruent sides.

4.

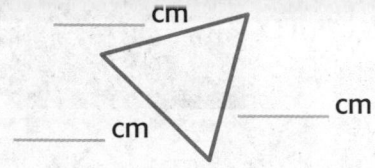

_____ cm

_____ cm

_____ cm

_____

_____

5.

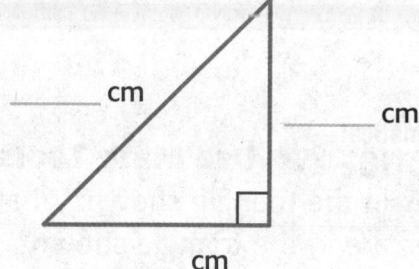

_____ cm

_____ cm

_____ cm

_____

_____

6.

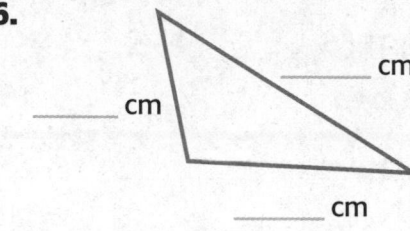

_____ cm

_____ cm

_____ cm

_____

_____

7.

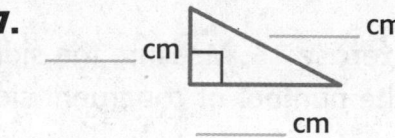

_____ cm

_____ cm

_____ cm

_____

_____

Measure the angles of each triangle to the nearest degree. Then describe the number of acute, right, or obtuse angles.

8.

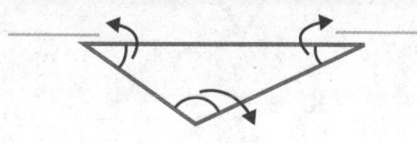

_____

_____

9.

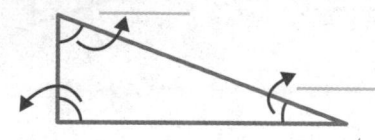

_____

_____

10.

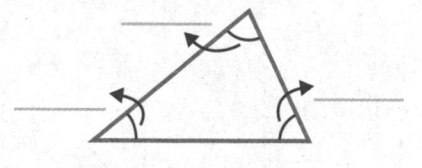

_____

_____

11.

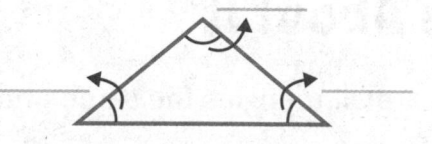

_____

_____

Apply It

12. In music, the "triangle" is an instrument with three congruent sides. If you know that the perimeter of the triangle is 18 inches, what is the measure of one side?

_____

13. Mathematical **PRACTICE** 5 **Use Math Tools** Measure the angles in the triangle shown. What type(s) of angles are in the triangle shown?

_____

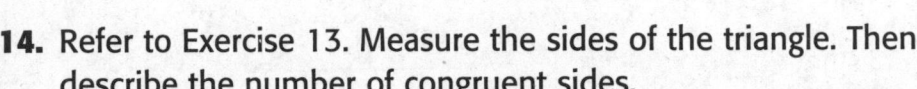

14. Refer to Exercise 13. Measure the sides of the triangle. Then describe the number of congruent sides.

_____

15. Mathematical **PRACTICE** 3 **Which One Doesn't Belong?** Circle the triangle that does not belong with the other three. Explain your reasoning.

  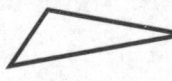

_____

_____

Write About It

16. How are all triangles the same and how can they be different?

_____

_____

MY Homework

Homework Helper

Need help? connectED.mcgraw-hill.com

Measure the sides of each triangle to the nearest tenth of a centimeter. Then describe the number of congruent sides.

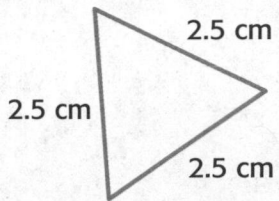

2.5 cm
2.5 cm
2.5 cm

The triangle has 3 congruent sides.

1.5 cm ◇ 1.5 cm
2.1 cm

The triangle has 2 congruent sides.

Measure the angles of each triangle to the nearest degree. Then describe the number of acute, right, or obtuse angles.

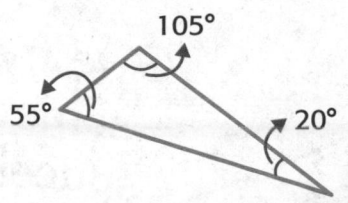

105°
55°
20°

The triangle has 1 obtuse angle and 2 acute angles.

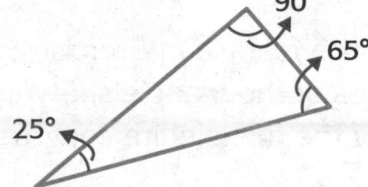

90°
65°
25°

The triangle has 1 right angle and 2 acute angles.

Practice

Measure the sides of each triangle to the nearest tenth of a centimeter. Then describe the number of congruent sides.

1. _____ cm _____ cm

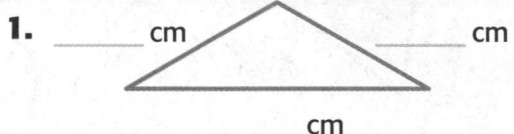

_____ cm

_____

_____

2.

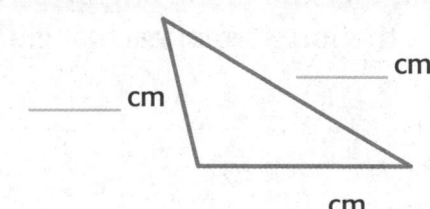

_____ cm

_____ cm

_____ cm

_____

_____

Cut and use this centimeter ruler. →

Measure the angles of each triangle to the nearest degree. Then describe the number of acute, right, or obtuse angles.

3.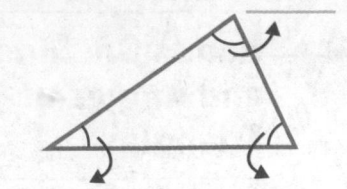

_____

_____

4.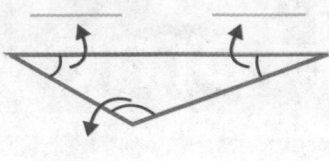

_____

_____

 Problem Solving

5. Measure the sides of the triangle shown. How many sides of the triangle are congruent?

_____

6. Refer to the triangle in Exercise 5. Measure the angles of the triangle shown. How many angles of the triangle are congruent?

_____

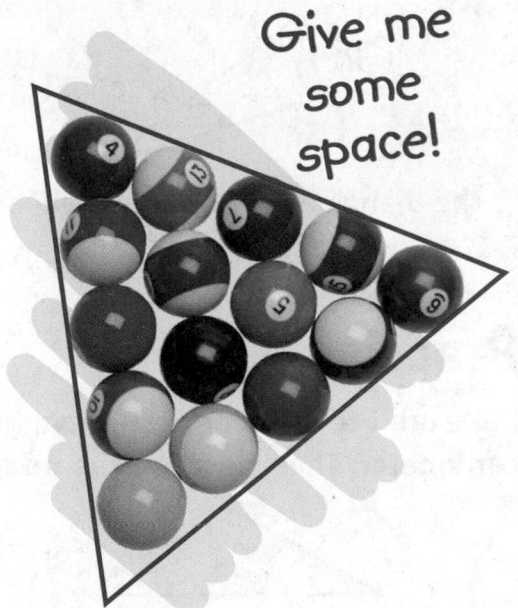

Give me some space!

7. In billiards, a rack is used to organize billiard balls at the beginning of the game. Jason is making the wood rack and found that each angle is congruent and that the sum of the angles is 180°. What is the measure of each angle?

_____

8. Measure each angle of the triangle. How many acute angles does the triangle have?

_____

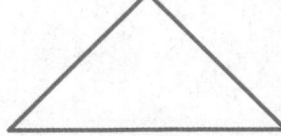

Classify Triangles

You can classify triangles using one or more of the following attributes. An **attribute** is a characteristic of a figure like side measures and angle measures.

 **Math in My World**  Watch  Tutor

Peanuts or pretzels?

Example 1

The Hammond family traveled from Columbus, Ohio, to Dallas, Texas, and then to Atlanta, Georgia, before returning home. The distance of each flight is shown on the map. Find the number of congruent sides.

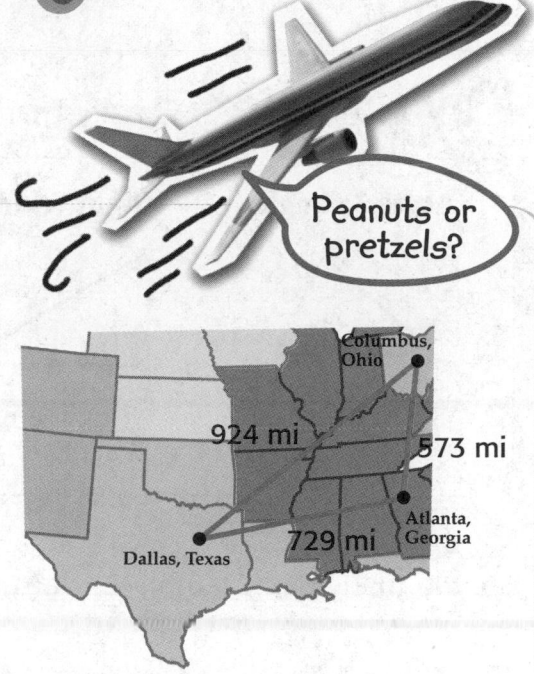

The lengths of the sides of the triangle are

924 miles, 573 miles, and _____ miles.

How many sides of the triangle are congruent? _____

Key Concept Classify Triangles by Sides

Isosceles Triangle	Equilateral Triangle	Scalene Triangle
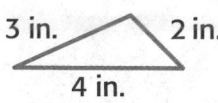 3 in. / 3 in. / 2 in.	3 in. / 3 in. / 3 in.	3 in. / 2 in. / 4 in.
at least two sides congruent	all sides congruent	no sides congruent

So, the triangle formed on the map in Example 1 is a

_____ triangle.

Example 2

Triangles form the sides of the Khafre Pyramid in Egypt. Determine the number of acute, obtuse, or right angles in the triangle.

How many angles of the triangle are acute? _____

How many angles of the triangle are obtuse? _____

How many angles of the triangle are right? _____

Key Concept   Classify Triangles by Angles

Acute Triangle	**Right Triangle**	**Obtuse Triangle**
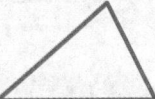	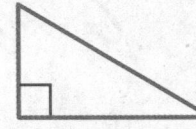	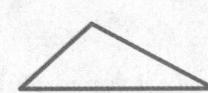
3 acute angles	1 right angle, 2 acute angles	1 obtuse angle, 2 acute angles

So, the triangle in Example 2 is a(n) _____ .

Guided Practice

1. Classify the triangle based on its sides.

How many sides of the triangle are congruent?

_____

The triangle is a(n) _____ .

2. Classify the triangle based on its angles.

The triangle is a(n) _____ .

Talk MATH

Describe an isosceles right triangle.

Independent Practice

Determine the number of congruent sides for each triangle. Then classify the triangle based on its sides.

3.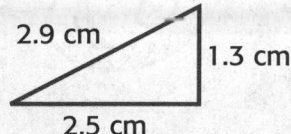
2.9 cm
1.3 cm
2.5 cm

_____

4.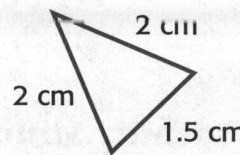
2 cm
2 cm
1.5 cm

_____

Classify each triangle based on its angles.

5.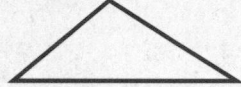

_____

6.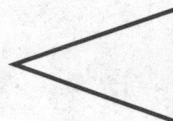

_____

7.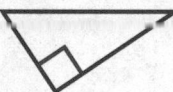

_____

8.

_____

Draw each triangle.

9. equilateral triangle

10. right triangle

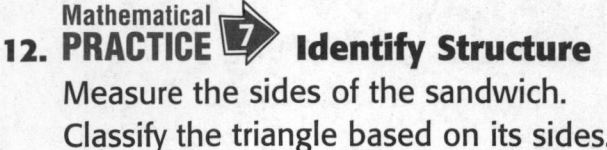

Problem Solving

11. Half of a rectangular sandwich looks like a triangle. Classify it based on its angles.

_____

12. **Identify Structure** Measure the sides of the sandwich. Classify the triangle based on its sides.

_____

Take a bite!

HOT Problems

13. **Draw a Conclusion** Emma, Gabriel, Jorge, and Makayla each drew a different triangle. Use the clues below to describe each person's triangle as isosceles, equilateral, or scalene and also as acute, right, or obtuse.

• Gabriel and Jorge each drew a 90° angle in their triangles.

• Gabriel's triangle does not have any congruent sides.

• One angle in Emma's triangle measures greater than 90°.

• Each side of Makayla's triangle and two sides of Emma's and Jorge's triangles are four centimeters long.

_____

_____

14. **Building on the Essential Question** How do I classify triangles using their attributes?

_____

_____

MY Homework

Homework Helper

Need help? ↗ connectED.mcgraw-hill.com

There is a large pyramid standing in front of the Louvre museum in Paris, France. The sides of the pyramid are shaped like triangles. Classify the red triangle based on its angles.

There are three acute angles.

So, the triangle formed by the side of the pyramid is an acute triangle.

Practice

1. Determine the number of congruent sides. Then classify the triangle based on its sides.

   How many sides of the triangle are congruent?

   _____

   The triangle is a _____.

   1.4 cm   3.1 cm
           2.7 cm

Vocabulary Check

Fill in each blank with the correct term(s) or number(s) to complete each sentence.

2. An equilateral triangle is a triangle with _____ congruent sides.

3. An acute triangle is a triangle with _____ angles each

   less than _____.

4. An obtuse triangle is a triangle with one angle that is greater

   than _____.

Problem Solving

5. Look at the triangle on the top of the White House in the photo. Describe the sides and angles of the triangle.

_____

_____

6. Serena has an art easel with sides of equal length. She opened the easel and placed it on her desk. Classify the type of triangle formed by the easel and the desk according to its sides. Next, classify the type of triangle formed by the easel and the desk according to its angles.

_____

7. Mathematical PRACTICE 7 Identify Structure The image shown at the right contains many triangles. Describe the different types of triangles found in the image.

_____

8. Mathematical PRACTICE 3 Justify Conclusions A triangle has two sides that are perpendicular. Could the triangle be isosceles, equilateral, or scalene? Explain.

_____

_____

_____

Test Practice

9. Which of the following figures is an obtuse triangle?

Ⓐ 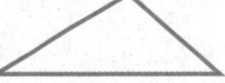　　Ⓑ 　　Ⓒ 　　Ⓓ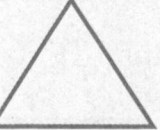

Check My Progress

Vocabulary Check

State whether each sentence is *true* or *false*.

1. A triangle with no congruent sides is a **scalene triangle**. _____

2. A polygon that has 4 sides and 4 angles is a **pentagon**. _____

3. Sides or angles with the same measure are **congruent**. _____

4. A **right triangle** is a triangle with two right angles. _____

Concept Check

Name each polygon. Determine if it appears to be *regular* or *not regular*.

5.

_____

6.

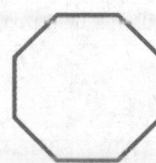

_____

Measure the sides of each triangle to the nearest tenth of a centimeter. Then describe the number of congruent sides.

7.

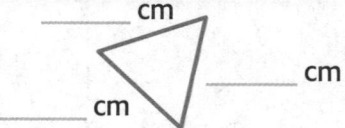

_____

_____

8.

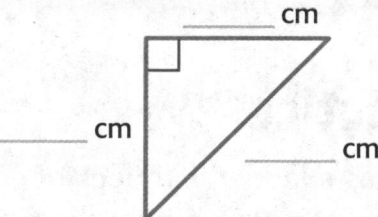

_____

_____

Problem Solving

9. Name the polygon shown by the video game screen at the right. Determine if it appears to be *regular* or *not regular*.

_____

10. Steve has three lengths of fence. He connects them to make a triangular pen for his dog. If the lengths are 5 meters, 6 meters, and 10 meters, what type of triangle is formed by the dog pen?

_____

11. Name the polygon shown by the banner at the right. Determine if it appears to be *regular* or *not regular*.

_____

12. Refer to the art in Exercise 11. Classify the triangle based on its angles.

_____

13. Lindsay was going to visit her grandmother, shop at the mall, and then return home. The route she took was in the shape of a triangle. The distance between each place she visited was 10 miles. What type of triangle is formed by the route she traveled?

_____

Test Practice

14. Miguel has a ladder with legs of equal length. He opened the ladder and placed it on the floor. What type of triangle is formed by the ladder and the floor?

   Ⓐ scalene triangle   Ⓒ equilateral triangle

   Ⓑ isosceles triangle   Ⓓ obtuse triangle

Hands On
Sides and Angles of Quadrilaterals

A quadrilateral is a polygon with four sides and four angles.

Measure It

Measure the sides and angles of each figure to determine if any are congruent. Then determine if any sides are parallel. Complete the table.

Figure 1   Figure 2   Figure 3   Figure 4

Attribute	Figure(s)
Opposite sides are congruent.	
Opposite sides are parallel.	
Opposite angles are congruent.	

Each figure has _____ sides and _____ angles.

Where has Polly gone?

CAFÉ

Talk About It

1. What common attributes do all of the figures have?

_____

2. Does Figure 3 have all the attributes of Figure 2? Explain.

_____

_____

Try It

Measure the sides and angles of each figure to determine if any are congruent. Then determine if any sides are parallel. Complete the table.

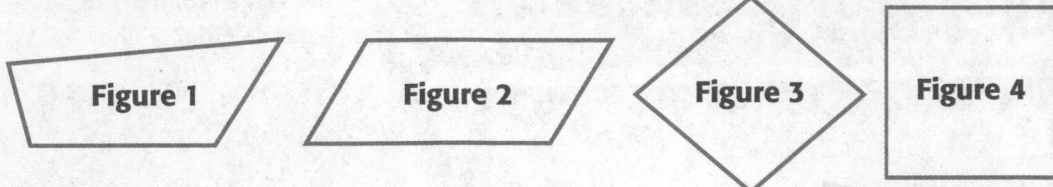

Figure 1   Figure 2   Figure 3   Figure 4

Attribute	Figure(s)
Opposite sides are congruent.	
Opposite sides are parallel.	
Opposite angles are congruent.	

Talk About It

3. Does Figure 3 have all the attributes of Figure 2? Explain.

_____

_____

4. What are some additional attributes that Figure 4 has that Figure 3 doesn't have?

_____

Mathematical
5. PRACTICE **1** **Make Sense of Problems** Explain how Figure 2 is a special kind of polygon.

_____

_____

_____

6. Which figure does not have any of the attributes listed in the table?

_____

Practice It

Measure the sides and angles of each figure to determine if any are congruent or parallel. Then answer Exercises 7–13.

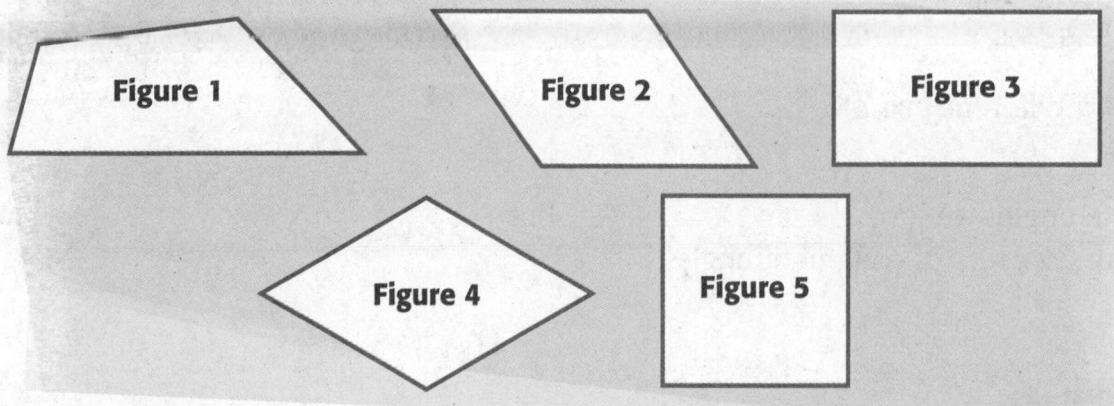

7. Complete the attributes of Figure 1.

   Opposite sides are _____ and _____ .

   Opposite angles are _____ .

   The figure has _____ sides and _____ angles.

8. Complete the attributes of Figure 2.

   Opposite sides are _____ and _____ .

   Opposite angles are _____ .

   The figure has _____ sides and _____ angles.

9. Which figures have all the attributes of Figure 1? _____

10. Which figures have all the attributes of Figure 2? _____

11. Which figures have all the attributes of Figure 3? _____

12. Which figures have four right angles? _____

13. Which figures have four equal sides? _____

14. Complete the attributes of the red quadrilateral outlining one side of the Chichen Itza pyramid in Mexico.

There is one pair of _____ opposite sides.

There is a different pair of _____ opposite sides.

Opposite angles are not _____, but there are two sets of congruent angles.

Mathematical
15. **PRACTICE** ▶2 **Reason** Explain one way to determine if a quadrilateral has parallel sides.

_____

_____

Mathematical
16. **PRACTICE** ▶3 **Which One Doesn't Belong?** Circle the quadrilateral that does not belong with the other three. Explain your reasoning.

_____

_____

Write About It

17. How are all quadrilaterals alike and how can they be different?

_____

_____

MY Homework

Homework Helper

Need help? connectED.mcgraw-hill.com

Measure the sides and angles of each figure to determine if any are congruent. Then determine if any sides are parallel. Complete the table.

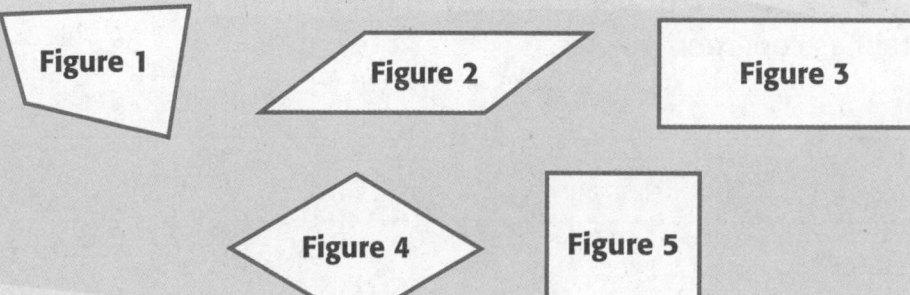

Attribute	Figure(s)
Opposite sides are congruent.	2, 3, 4, 5
Opposite sides are parallel.	2, 3, 4, 5
Opposite angles are congruent.	2, 3, 4, 5

Each figure has 4 sides and 4 angles.

Practice

Refer to the figures above in the Homework Helper to solve Exercises 1–3.

1. Complete the attributes of Figure 2.

   Opposite sides are _____ and _____.

   Opposite angles are _____.

   The figure has _____ sides and _____ angles.

2. Which figures have all the attributes of Figure 2? _____

3. Which figures have four right angles? _____

Problem Solving

4. The state of Nevada is in the shape of a quadrilateral.
Complete the attributes of the outline of the state of Nevada.

There is one set of _____ opposite sides.

Opposite sides are not _____.

Opposite angles are not _____,
but there are two right angles.

Mathematical
5. PRACTICE 2 Reason Explain one way to determine if a
quadrilateral has congruent angles.

_____

_____

Mathematical
6. PRACTICE 3 Which One Doesn't Belong? Circle the
quadrilateral that does not belong with the other three. Explain
your reasoning.

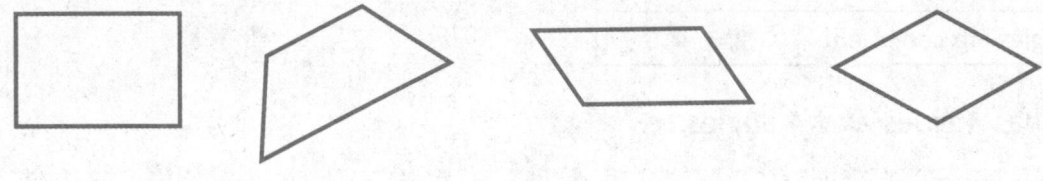

_____

_____

Vocabulary Check

**Fill in each blank with the correct term or number to complete
the sentence.**

7. A quadrilateral is a polygon with _____ sides and _____ angles.

Classify Quadrilaterals

Lesson 5

ESSENTIAL QUESTION
How does geometry help me solve problems in everyday life?

You can classify quadrilaterals using one or more of the following attributes like congruent sides, parallel sides, and right angles.

 ## Math in My World ▶Watch

Example 1

Trina cut out polygon mats to use for her travel photos. Use the figures below to determine the missing attribute(s) of each type of quadrilateral.

Quadrilateral

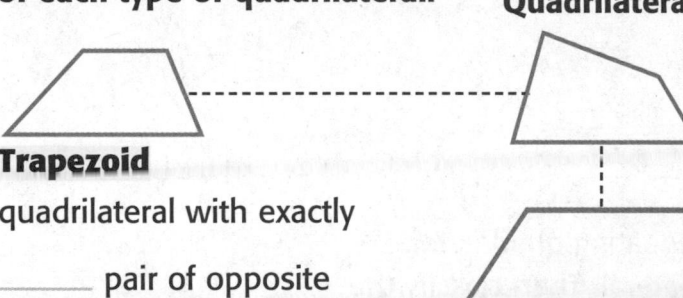

Trapezoid
quadrilateral with exactly

_____ pair of opposite sides parallel

Parallelogram
quadrilateral with opposite sides congruent and

_____

Rectangle
parallelogram with

_____ right angles

Square
parallelogram with

_____ sides congruent

and _____ right angles

Rhombus
parallelogram with

_____ sides congruent

A square has all the attributes of a rectangle and a _____.

Example 2

One side of the Realia building in Madrid, Spain, is shown at the right. Describe the attributes of the quadrilateral. Then classify it based on its attributes.

The quadrilateral has opposite sides _____

and _____ .

So, it is a _____ .

Guided Practice

1. Describe the attributes of the quadrilateral below. Then classify the quadrilateral based on its attributes.

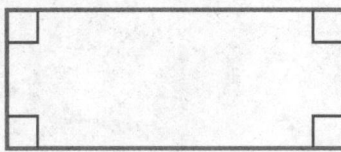

The opposite sides of the quadrilateral are _____

and _____ .

There are _____ right angles.

So, the quadrilateral is a _____ .

2. The design below is made up of a repeating quadrilateral. Describe the attributes of the quadrilateral. Then classify the quadrilateral based on its attributes.

The quadrilateral has _____ congruent sides.

Opposite sides are _____ .

So, the quadrilateral is a _____ .

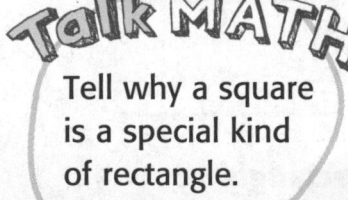

Tell why a square is a special kind of rectangle.

Independent Practice

Describe the attributes of each quadrilateral. Then classify the quadrilateral.

3.

_____

_____

_____

4.

_____

_____

_____

5. Circle the quadrilateral(s) that have all the attributes of a parallelogram.

rectangle   rhombus   square   trapezoid

6. Circle the quadrilateral(s) that have all the attributes of a rhombus.

rectangle   square   trapezoid   parallelogram

State whether the following statements are _true_ or _false_. If _false_, explain why.

7. All parallelograms have opposite sides congruent and parallel. Since rectangles are parallelograms, all rectangles have opposite sides congruent and parallel.

_____

_____

8. All squares have four congruent sides. Since rectangles are squares, all rectangles have four congruent sides.

_____

_____

Problem Solving

9. Mathematical PRACTICE 7 Identify Structure Many aircraft display the shape of the American flag as shown below to indicate motion. Classify the quadrilateral.

_____

10. Adena used a quadrilateral in her art design. The quadrilateral has no sides congruent and only one pair of opposite sides parallel. Classify the shape of the quadrilateral she used.

_____

11. Traci planted two tomato gardens. One garden is rectangular. The shape of the second garden has all the attributes of the rectangular garden. In addition, it has four congruent sides. Classify the shape of the second tomato garden.

One smart tomato!

_____

HOT Problems

12. Mathematical PRACTICE 4 Model Math Draw a parallelogram that is neither a square, rhombus, nor rectangle.

13. ❓ Building on the Essential Question How do I classify quadrilaterals using their attributes?

_____

_____

MY Homework

Homework Helper

Need help? connectED.mcgraw-hill.com

Describe the attributes of the quadrilateral. Then classify it based on its attributes.

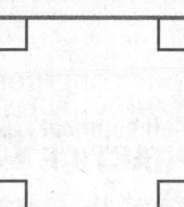

The quadrilateral has all sides congruent and opposite sides parallel.

It has four right angles.

So, the quadrilateral is a square.

Practice

Describe the attributes of each quadrilateral. Then classify the quadrilateral.

1.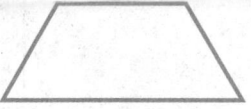

_____

_____

_____

2.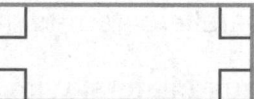

_____

_____

_____

3. Circle the quadrilateral(s) that have all the attributes of a rectangle.

    trapezoid   parallelogram   square   rhombus

Problem Solving

Name all the quadrilaterals that have the given attributes.

4. opposite sides parallel

_____

5. four right angles _____

6. exactly one pair of opposite sides parallel _____

7. four congruent sides _____

8. **Mathematical**
PRACTICE **Model Math** Write a real-world problem that
involves classifying a quadrilateral. Then solve the problem.

_____

_____

Vocabulary Check

**Fill in each blank with the correct term or number to complete
each sentence.**

9. A rectangle is a parallelogram with _____ right angles.

10. A trapezoid is a quadrilateral with exactly _____ pair of
parallel sides.

Test Practice

11. Which statement about the figures shown below is true?

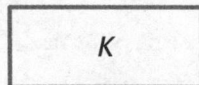  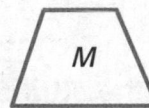 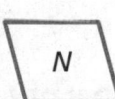

    Ⓐ Figures *K* and *N* are rectangles.

    Ⓑ Figures *L* and *N* are quadrilaterals.

    Ⓒ Figures *K* and *N* are parallelograms.

    Ⓓ Figures *M* and *N* are parallelograms.

Name ....

Hands On
Build Three-Dimensional Figures

Lesson 6

ESSENTIAL QUESTION
How does geometry help me solve problems in everyday life?

A **three-dimensional figure** has length, width, and height. A **net** is a two-dimensional pattern of a three-dimensional figure. You can use a net to build a three-dimensional figure.

A **cube** is a three-dimensional figure with six faces that are congruent squares. **Congruent figures** have the same size and shape.

A **rectangular prism** is a three-dimensional figure with six rectangular faces. Opposite faces are parallel and congruent.

A **face** is a flat surface.

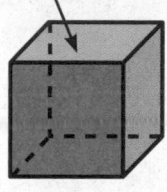

Cube

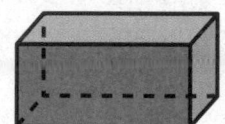

Rectangular Prism

Build It

 Copy the net shown onto grid paper.

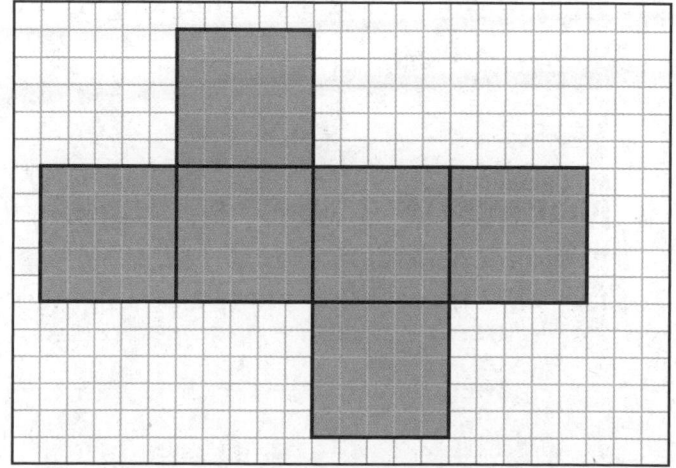

Cut out the net. Fold along the lines to form a three-dimensional figure. What figure did you form?

_____

Try It

 Copy the net shown onto grid paper.

 Cut out the net. Fold along the lines to form a three-dimensional figure. What figure did you form?

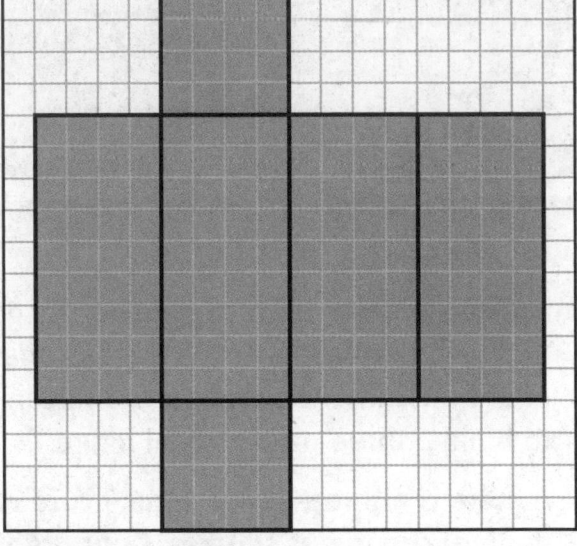

_____

How are the two figures you just built alike?

_____

_____

How are the two figures you just built different?

_____

Talk About It

1. In the first activity, what two-dimensional figure forms the faces of the figure? How many faces are there? How many are congruent?

_____

2. Identify the length, width, and height of the cube you formed in the first activity.

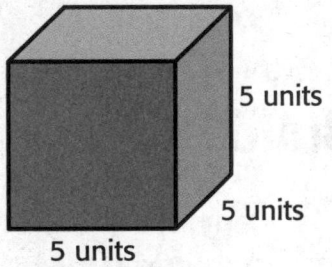

5 units

5 units

5 units

_____

3. What do you notice about the length, width, and height of the cube?

4. **Mathematical PRACTICE 7 Identify Structure** In the second activity, what two-dimensional figures form the faces of the figure? How many faces are there? How many are congruent?

_____

Name ....

Practice It

For Exercises 5 and 6, refer to the grid at the right.

5. Copy the net onto grid paper. Cut out the net and fold along the lines to form a three-dimensional figure. What figure did you form?

_____

6. What two-dimensional figure forms the faces of the figure?

_____

How many faces are there? _____ Describe the congruent faces.

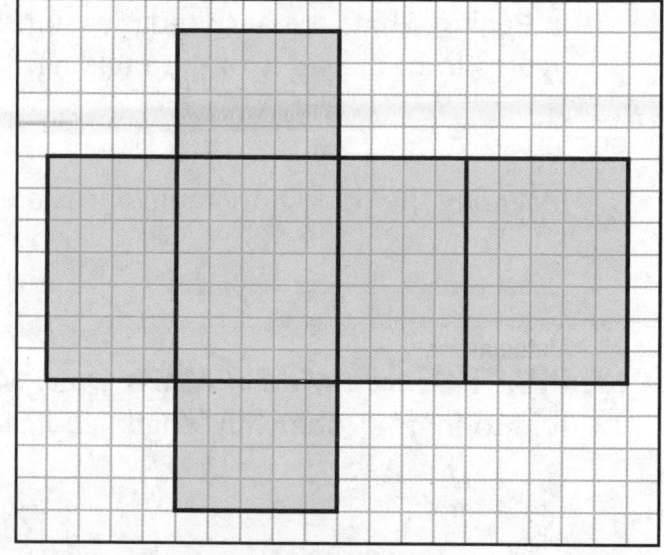

For Exercises 7–9, refer to the grid at the right.

7. Copy the net onto grid paper. Cut out the net and fold along the lines to form a three-dimensional figure. What figure did you form?

_____

8. What two-dimensional figure forms the faces of the figure?

_____

How many faces are there?

_____

Describe the congruent faces.

_____

9. Identify the length, width, and height of the figure you formed.

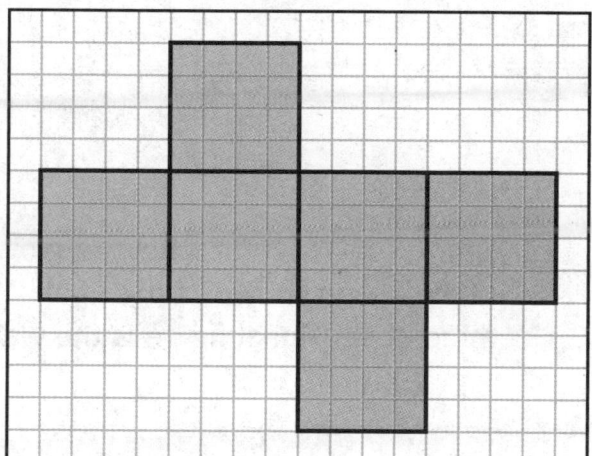

Apply It

10. The rectangular prism-shaped building shown at the right was used for the 2008 Olympics in Beijing, China. What two-dimensional figures form the sides of the building?

_____

Including the floor, how many faces are there?

_____

11. Mathematical **PRACTICE** 4 **Model Math** Draw two different nets that would fold to form a cube with length, width, and height each 4 units.

My Drawing!

12. Farmers have learned how to grow watermelons in the shape shown at the right. What three-dimensional figure is the watermelon?

What happened to me?

_____

Write About It

13. How are nets used to build three-dimensional figures?

_____

_____

_____

MY Homework

Homework Helper

Need help? ⟋ connectED.mcgraw-hill.com

The net shown was used to form the three-dimensional figure below.

The three-dimensional figure formed from the net is a rectangular prism.

The faces of the rectangular prism are rectangles.

The figure has 6 faces.

The four rectangles are congruent, and the two squares are congruent.

The figure formed has a length of 4 units, a width of 6 units, and a height of 6 units.

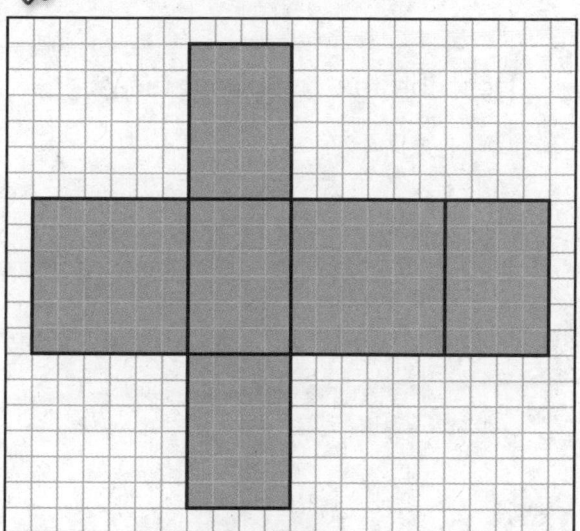

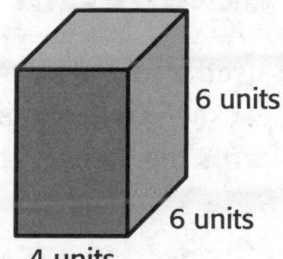

6 units

6 units

4 units

Vocabulary Check

Fill in each blank with the correct word(s) to complete each sentence.

1. A three-dimensional figure has _____, width,

   and _____.

2. A net is a two-dimensional _____ of a three-dimensional figure.

3. A cube is a three-dimensional figure with six square faces

   that are _____.

Practice

For Exercises 4–6, refer to the grid at the right.

4. What three-dimensional figure is formed using the net shown?

   _____

5. What two-dimensional figure forms the sides of the figure?

   _____

   Describe the congruent faces.

   _____

   _____

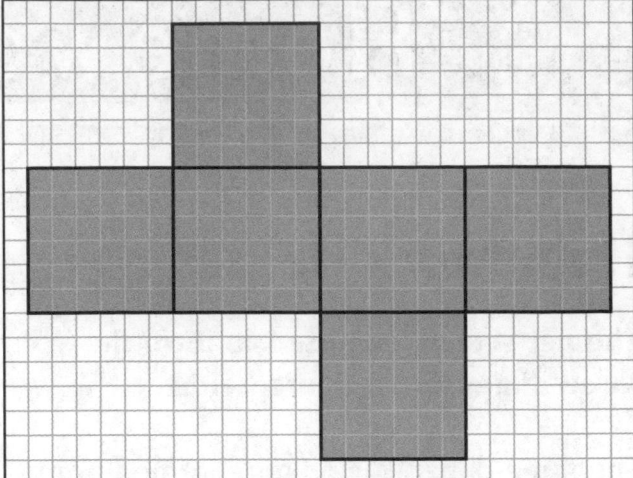

6. Identify the length, width, and height of the figure formed.

   _____

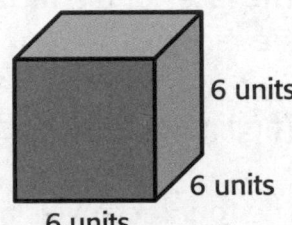

6 units

6 units

6 units

Problem Solving

7. Rachel used a rectangular prism-shaped box to ship a package to her friend. What two-dimensional figure forms the faces of the box?

   _____

   Including the bottom, how many faces are there?

   _____

   Describe the faces.

   _____

   _____

SPECIAL DELIVERY

8. Joseph is forming a three-dimensional figure using a net. The figure has six congruent square faces. What type of figure did he make?

   _____

Three-Dimensional Figures

Lesson 7

ESSENTIAL QUESTION
How does geometry help me solve problems in everyday life?

A **three-dimensional figure** has length, width, and height.

A **face** is a flat surface.

A **vertex** is a point where 3 or more faces meet.

An **edge** is where two faces meet.

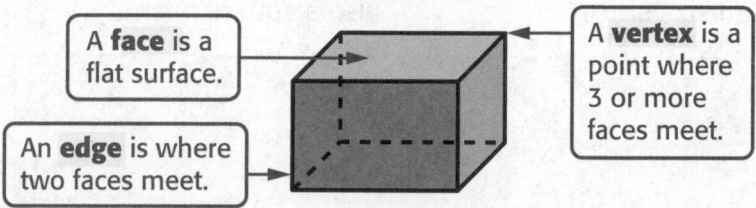

Math in My World

  Tools Watch Tutor

Describe the faces, edges, and vertices of the figure outlined on the luggage bag. Then identify the shape of the figure.

faces The figure has _____ faces. Each face appears to be a rectangle.

edges There are _____ edges. The opposite edges are parallel and congruent.

vertices The figure has _____ vertices.

Prisms are three-dimensional figures. A **prism** has at least three faces that are rectangles. The top and bottom faces, called the **bases**, are congruent parallel polygons.

The figure above is a rectangular prism. In a **rectangular prism**, the bases are congruent rectangles. A rectangular prism has six rectangular faces, twelve edges, and eight vertices.

Key Concept Prisms

Rectangular Prism	Triangular Prism	Cube
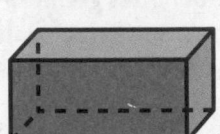	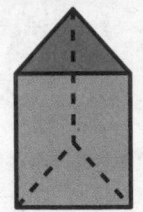	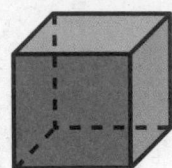
A rectangular prism has six rectangular faces, twelve edges, and eight vertices.	A triangular prism has triangular bases. It has five faces, nine edges, and six vertices.	A cube has six square faces, twelve edges, and eight vertices. A cube is also a square prism.

Guided Practice Check ☑

1. Describe the faces, edges, and vertices of the three-dimensional figure. Then identify it.

faces This figure has _____ faces. The _____ bases are congruent and parallel. The other faces are _____.

edges There are _____ edges. The edges that form the vertical sides of the rectangles are parallel and _____.

vertices This figure has _____ vertices.

The figure is a _____.

Talk MATH
Describe the differences between a triangular prism and a rectangular prism.

Independent Practice

Describe the faces, edges, and vertices of each three-dimensional figure. Then identify it.

2.
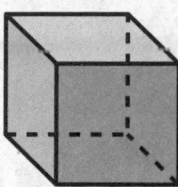

_____

_____

_____

3.
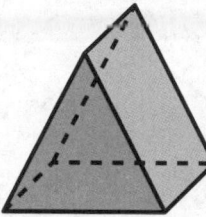

_____

_____

_____

4.
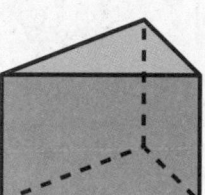

_____

_____

_____

5.
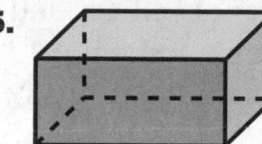

_____

_____

_____

6.

_____

_____

_____

7.
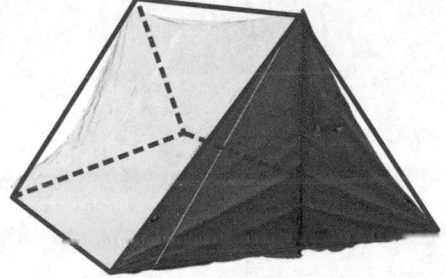

_____

_____

_____

Problem Solving

Real World

8. Mathematical **PRACTICE 7 Identify Structure** The Metropolitan Correction Center in Chicago is in the shape of a triangular prism. Circle the two-dimensional figures that make up the faces of the prism.

9. Describe the number of vertices and edges in an unopened cereal box. Identify the shape of the box.

HOT Problems

10. Mathematical **PRACTICE 4 Model Math** What figure is formed if only the height of a cube is increased? Draw the figure to support your answer.

My Drawing!

11. **Building on the Essential Question** Why is it important to know the different properties of three-dimensional figures?

Name ....

MY Homework

Homework Helper

Need help? connectED.mcgraw-hill.com

Describe the faces, edges, and vertices of the ramp. Then identify the shape of the ramp.

faces   This figure has 5 faces. The triangular bases are congruent and parallel. The other faces are rectangles.

edges   There are 9 edges. The edges that form the horizontal sides of the rectangles are parallel and congruent.

vertices   This figure has 6 vertices.

The ramp is a triangular prism.

Practice

Describe the faces, edges, and vertices of each three-dimensional figure. Then identify it.

1.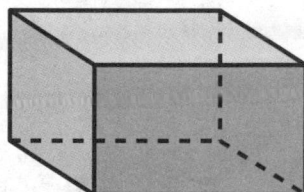

_____

_____

_____

2.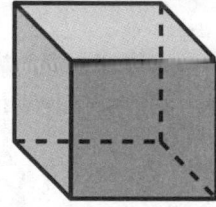

_____

_____

_____

Problem Solving

3. Rhett made a simple drawing of his house. It is a three-dimensional figure with four faces that are rectangular and two that are square. What kind of figure is it?

_____

4. A toy box has 6 faces that are squares. There are 12 edges and 8 vertices. Identify the shape of the toy box.

_____

Mathematical
5. PRACTICE **Make Sense of Problems** Gabriel is playing a board game. When it is his turn, he tosses a three-dimensional figure that has 6 square faces. What kind of figure is it? How many edges and vertices does it have?

_____

Vocabulary Check

Fill in the blank with the correct term or number to complete the sentence.

6. A vertex is a point where _____ or more edges meet.

Test Practice

7. Which statement is true about the three-dimensional figure that most closely represents the slice of pie?

Ⓐ The figure has 4 vertices.

Ⓑ The figure has 6 vertices.

Ⓒ The figure has 8 vertices.

Ⓓ The figure has 9 vertices.

Easy as pie

Need more practice? Download Extra Practice at ⚡ **connectED.mcgraw-hill.com**

Check My Progress

Vocabulary Check

Circle the correct term or terms to complete each sentence.

1. A (**rectangular prism**, **triangular prism**) is a three-dimensional figure that has six rectangular faces and eight vertices.

2. A (**rectangle**, **rhombus**) is a parallelogram with all sides congruent.

3. A(n) (**vertex**, **edge**) of a three-dimensional figure is where two faces meet.

4. A (**prism**, **trapezoid**) has at least three faces that are rectangles.

Concept Check

Describe the attributes of each quadrilateral. Then classify the quadrilateral.

5.

_____

_____

_____

6.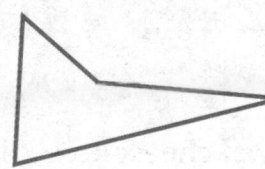

_____

_____

_____

7. Circle the quadrilateral(s) that have all the attributes of a rhombus.

   trapezoid   square   parallelogram   rectangle

8. Circle the quadrilateral(s) that have all the attributes of a rectangle.

   parallelogram   square   trapezoid   rhombus

Describe the faces, edges, and vertices of each three-dimensional figure. Then identify it.

9.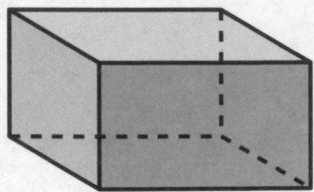

_____

_____

_____

10.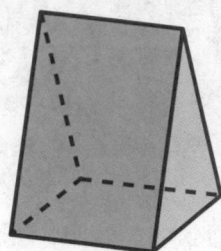

_____

_____

_____

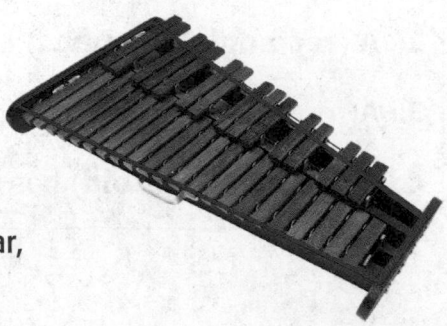

Problem Solving

11. Kendra's xylophone is in the shape of the trapezoid shown at the right. Describe any parallel, perpendicular, or congruent sides of the xylophone.

_____

_____

_____

12. Garrett cut a piece of cheese to eat for a snack. The cheese is a prism with 3 faces that are rectangular and 2 that are triangular. What kind of figure is it?

_____

Test Practice

13. Which is *not* a true statement?

Ⓐ All squares are parallelograms.

Ⓑ Some rhombi are squares.

Ⓒ All rectangles are squares.

Ⓓ All rectangles are parallelograms.

Hands On
Use Models to Find Volume

Lesson 8

ESSENTIAL QUESTION
How does geometry help
me solve problems in
everyday life?

Volume is the amount of space inside a three-dimensional figure. Centimeter cubes can help you find the volume of a prism.

What's Up?

Build It

Use centimeter cubes to build four different rectangular prisms. Complete the fourth and fifth columns of the table below for each prism.

Prism	Length (cm)	Width (cm)	Height (cm)	Number of Cubes	Volume (cubic cm)
A	1	2	1		
B	2	2	1		
C	3	2	2		
D	4	2	2		

A prism built from cubes has no gaps or overlaps.

A cube with a side length of one unit is called a **unit cube.**
A unit cube has a volume of 1 cubic unit, or 1 unit³.
A **cubic unit** is a unit for measuring volume.

 1 cubic unit    2 cubic units   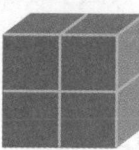 4 cubic units

So, if you use 12 centimeter cubes to build a rectangular prism, the prism has a

volume of _____ cubic centimeters, or _____ cm³.

Try It

Use centimeter cubes to build the rectangular prism shown. Complete the table for each layer.

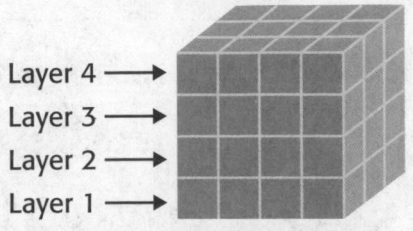

Layer 4 →
Layer 3 →
Layer 2 →
Layer 1 →

Layer	Length (cm)	Width (cm)	Height (cm)	Number of Cubes	Volume (cubic cm)
1					
2					
3					
4					

How many cubes were used to build the prism? _____

What is the volume? _____

Talk About It

1. Describe the relationship between the number of cubes needed to build a rectangular prism and its volume, in cubic units.

   _____

   _____

2. Describe the pattern in the table between the length, width, height, and volume of each prism.

   _____

   _____

3. Use ℓ for length, w for width, and h for height to write a formula for the volume V of a rectangular prism.

   _____

4. **Mathematical PRACTICE 5** **Use Math Tools** Use your formula to find the volume of the prism at the right in appropriate units. Verify your solution by counting the number of cubes.

   _____

Practice It

Mathematical PRACTICE 5 **Use Math Tools** Use centimeter cubes to build the rectangular prism shown.

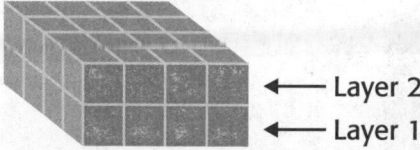

← Layer 2
← Layer 1

5. Complete the table below.

Layer	Length (cm)	Width (cm)	Height (cm)	Number of Cubes
1				
2				

6. How many cubes were used to build the prism? _____

What is the volume? _____ cm³

Use centimeter cubes to build the rectangular prism shown.

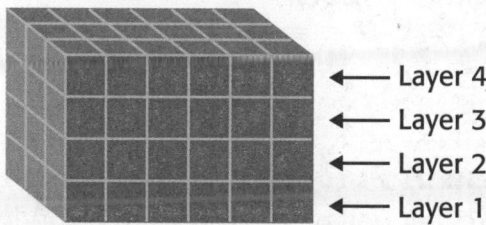

← Layer 4
← Layer 3
← Layer 2
← Layer 1

7. Complete the table below.

Layer	Length (cm)	Width (cm)	Height (cm)	Number of Cubes	Volume (cubic cm)
1					
2					
3					
4					

8. How many cubes were used to build the prism? _____

What is the volume? _____ cm³

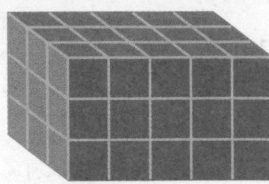

Apply It

Use the prism shown for Exercises 9–11.

9. What shape is the base of the prism?

_____

Mathematical
10. PRACTICE ⑥ **Explain to a Friend** Explain to a friend how to find the area of the base of the prism.

_____

_____

11. Find the volume of the prism above by multiplying the area of the base by the height. Verify your solution by counting the number of centimeter cubes.

_____

Mathematical
12. PRACTICE ① **Make Sense of Problems** Valerie knows that the volume of a prism is 36 cubic units. She knows that the length of the prism is 4 units and the width is 3 units. What is the height of the prism?

_____

Write About It

13. Describe a way to find the volume of a rectangular prism without using models.

_____

_____

Name ....

MY Homework

Homework Helper

Need help? connectED.mcgraw-hill.com

Centimeter cubes were used to build the rectangular prism shown. The table shows the number of cubes needed to build each layer.

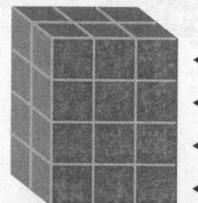

← Layer 4
← Layer 3
← Layer 2
← Layer 1

Layer	Length (cm)	Width (cm)	Height (cm)	Number of Cubes	Volume (cubic cm)
1	3	2	1	6	6
2	3	2	1	6	6
3	3	2	1	6	6
4	3	2	1	6	6

So, 24 cubes were used to build the prism.

The volume of the prism is 24 cubic centimeters, or 24 cm³.

Vocabulary Check

Fill in each blank with the correct term or number to complete each sentence.

1. Volume is the amount of _____ inside a three-dimensional figure.

2. A cube with a side length of _____ unit is called a unit cube.

3. The volume of a rectangular prism can be found by multiplying the

length by the _____ by the height.

Practice

For Exercises 4-7, centimeter cubes were
used to build the rectangular prism shown.

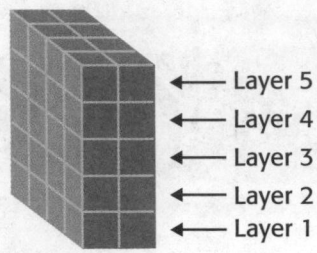

← Layer 5
← Layer 4
← Layer 3
← Layer 2
← Layer 1

4. How many cubes were needed to build Layer 1?

_____

5. Complete the table below.

Layer	Length (cm)	Width (cm)	Height (cm)	Number of Cubes	Volume (cubic cm)
1					
2					
3					
4					
5					

6. How many cubes were used to build the prism?

_____

7. What is the volume of the prism?

_____

Problem Solving

8. **Mathematical PRACTICE** **1** **Make Sense of Problems** Samir knows that the volume of a prism is 40 cubic units. He also knows that the width of the prism is 2 units and the height is 5 units. What is the length of the prism?

_____

9. Centimeter cubes were used to build the prism. What is the volume of the prism?

_____

Volume of Prisms

Lesson 9

ESSENTIAL QUESTION
How does geometry help me solve problems in everyday life?

Volume is the amount of space inside a three-dimensional figure. You can use either formula below to find the volume of a prism.

$V = \ell \times w \times h$ V = volume, ℓ = length, w = width, and h = height

$$B = \ell w$$

$V = B \times h$ V = volume, B = area of the base, and h = height

Common units of volume are cubic inches, cubic feet, cubic yards, cubic centimeters, and cubic meters.

Math in My World

Example 1

On his family vacation to the beach, Armando filled a cooler with water and snacks. Find the volume of the cooler.

Prisms are cooler!

20 in.

15 in.

30 in.

One Way Use $V = \ell \times w \times h$.

$V = \ell \times w \times h$ Volume formula

$V = $ _____ $\times$ _____ $\times$ _____ $\ell = 30, w = 15, h = 20$

$V = $ _____   Multiply.

Another Way Use $V = B \times h$.

$V = B \times h$ Volume formula

$V = $ _____ $\times$ _____ $B = 30 \times 15, h = 20$

$V = $ _____   Multiply.

The volume of the cooler is _____ cubic inches.

Remember that the Associative Property of Multiplication tells you that the way in which factors are grouped does not change the product. You can group the factors to make the multiplication easier.

Example 2

Find the volume of the prism.

$V = \ell \times w \times h$ Volume formula

$V = 17 \times 7 \times 9$

$V = \underline{\hspace{1cm}} \times (\underline{\hspace{1cm}} \times \underline{\hspace{1cm}})$ Associative Property

$V = \underline{\hspace{1cm}} \times \underline{\hspace{1cm}}$ Multiply.

$V = \underline{\hspace{1cm}}$ Multiply.

The volume of the prism is $\underline{\hspace{1cm}}$ cm³.

9 cm

7 cm

17 cm

Guided Practice ☑

Find the volume of each prism.

1.

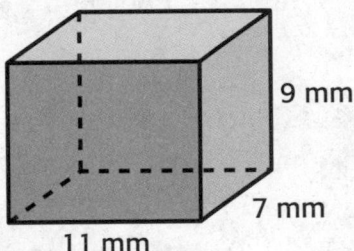

9 mm

7 mm

11 mm

$V = \ell \times w \times h$

$V = \underline{\hspace{1cm}} \times \underline{\hspace{1cm}} \times \underline{\hspace{1cm}}$

$V = \underline{\hspace{1cm}}$ mm³

2.

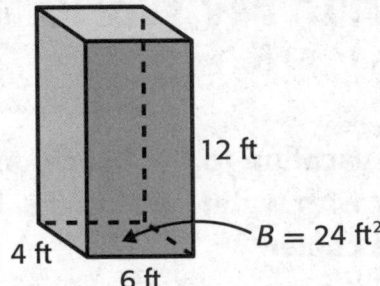

12 ft

$B = 24$ ft²

4 ft

6 ft

$V = B \times h$

$V = \underline{\hspace{1cm}} \times \underline{\hspace{1cm}}$

$V = \underline{\hspace{1cm}}$ ft³

Talk MATH

If you know the area of the base of a rectangular prism and the prism's height, which formula would you use? Why?

Independent Practice

Mathematical PRACTICE 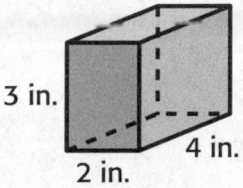 **2** **Use Symbols** Find the volume of each prism. Use the formula $V = \ell \times w \times h$ or $V = B \times h$.

3.

3 in.

4 in.

2 in.

$V =$ _____

4.

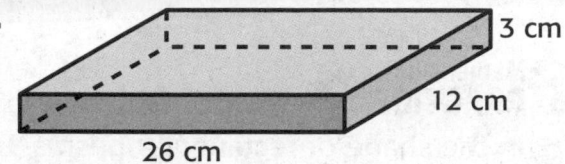

3 cm

12 cm

26 cm

$V =$ _____

5.

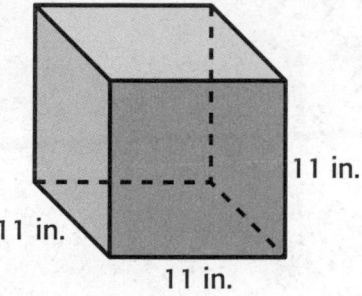

11 in.

11 in.

11 in.

$V =$ _____

6. 16 m

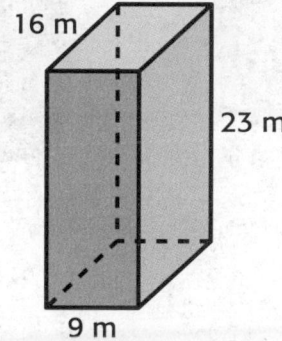

23 m

9 m

$V =$ _____

7.

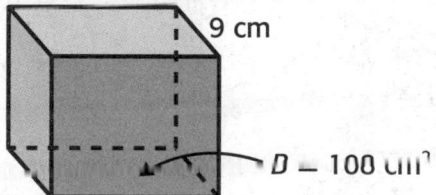

9 cm

$B = 108 \text{ cm}^2$

$V =$ _____

8.

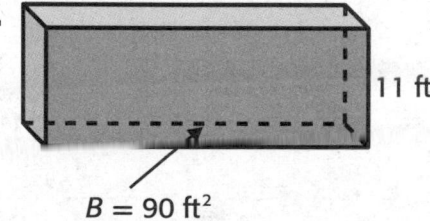

11 ft

$B = 90 \text{ ft}^2$

$V =$ _____

Problem Solving

9. Find the volume of the Frog Queen building in Graz, Austria. The building is 18 meters long, 17 meters tall, and 18 meters wide.

_____

10. **Mathematical PRACTICE 4** **Model Math** Two pet carriers are in the shape of rectangular prisms. Determine the volume of each pet carrier. Circle the carrier that has the greater volume.

Land Carrier: _____ in³ Olympic Carrier: _____ in³

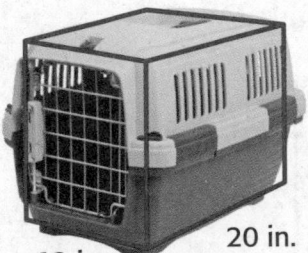

12 in.

20 in.

12 in.

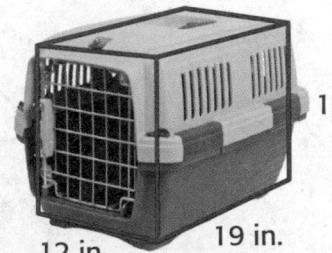

10 in.

12 in.

19 in.

HOT Problems

11. **Mathematical PRACTICE 2** **Use Number Sense** Explain how the Associative Property can be used to mentally find the volume of the prism shown.

_____

_____

_____

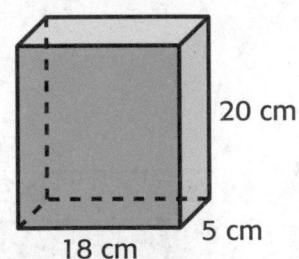

20 cm

18 cm

5 cm

12. **Building on the Essential Question** How do I find the volume of rectangular prisms?

_____

Name _____

MY Homework

Lesson 9

Volume of Prisms

Homework Helper

Need help? 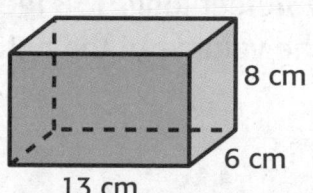 connectED.mcgraw-hill.com

Find the volume of the prism.

$V = \ell \times w \times h$ Volume formula

$V = 13 \times 6 \times 8$ $\ell = 13, w = 6, h = 8$

$V = 624$ Multiply.

The volume of the prism is 624 cm³.

8 cm
6 cm
13 cm

Practice

Find the volume of each prism.

1.

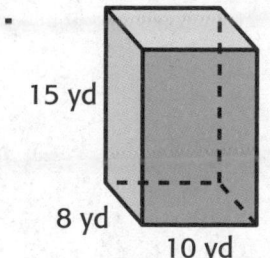

15 yd

8 yd

10 yd

$V =$ _____

2.

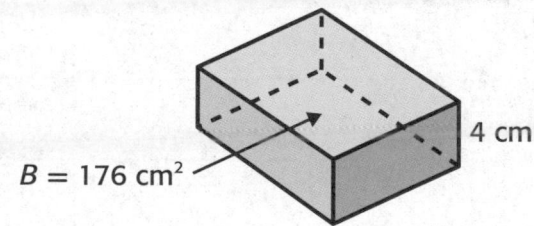

4 cm

$B = 176\ cm^2$

$V =$ _____

Vocabulary Check

Fill in the blank with the correct term or number to complete the sentence.

3. Volume is measured in _____ units.

Problem Solving

4. The Donaldson's swimming pool measures 15 meters long, 8 meters wide, and 3 meters deep. How many cubic meters of water will the pool hold?

_____

5. The hotel that the Hutching family is staying at on vacation is shaped like a rectangular prism. It is 234 feet long, 158 feet wide, and 37 feet tall. What is the volume of the hotel?

_____

6. Jena keeps her recipes in a box with dimensions 7 inches, 5 inches, and 4 inches. What is the volume of the box?

_____

7. **Mathematical PRACTICE 4 Model Math** Describe the dimensions of two different prisms that each have a volume of 2,400 cubic centimeters. Then draw each prism.

_____

_____

_____

My Drawing!

Test Practice

8. What is the volume of the prism formed by the luggage bag?

Ⓐ 6,000 in³

Ⓑ 6,600 in³

Ⓒ 7,200 in³

Ⓓ 7,400 in³

22 in.

30 in.

10 in.

Name ....

Hands On
Build Composite Figures

ESSENTIAL QUESTION
How does geometry help me solve problems in everyday life?

A **composite figure** is made up of two or more three-dimensional figures.

Build It

A composite figure is shown below. Use centimeter cubes to build the figure.

1 Count the number of cubes needed to make the base layer.

How many cubes did you use? ⟶ ☐

2 Count the number of cubes needed to make the top layer.

How many cubes did you use? ⟶ + ☐

3 Add the number of cubes for the base and the top. ⟶ ☐

Talk About It

1. How many cubes did it take to build the figure? _____

2. What is the volume of the composite figure?

_____ cubic centimeters

Try It

**Separate the composite figure into two rectangular prisms.
Then find the volume of each prism.**

 Find the volume of the top prism.

$V = \ell \times w \times h$

$V = \underline{\hspace{1cm}} \times \underline{\hspace{1cm}} \times \underline{\hspace{1cm}}$

$V = \underline{\hspace{1cm}}$

The volume of the top prism is _____ cubic centimeters.

 Find the volume of the bottom prism.

$V = \ell \times w \times h$

$V = \underline{\hspace{1cm}} \times \underline{\hspace{1cm}} \times \underline{\hspace{1cm}}$

$V = \underline{\hspace{1cm}}$

The volume of the bottom prism is _____ cubic centimeters.

 Add the volumes to find the volume of the composite figure.

_____ + _____ = _____

So, the volume of the composite figure is _____ cubic centimeters.

Talk About It

3. Explain how you can use addition to find the volume of a
composite figure.

_____

_____

4. Mathematical
PRACTICE **1** **Make Sense of Problems** Explain how you
would find the volume of the composite figure shown.

_____

_____

5. What is the volume of the figure in Exercise 4?

_____ cubic centimeters

Practice It

Use the model at the right to build the composite figure using centimeter cubes.

6. Separate the figure into prisms. Make a drawing of each prism used to build the composite figure.

My Drawing!

7. How many cubes did it take to build the figure?

_____

8. What is the volume of this figure?

_____ cubic centimeters

Use the model at the right to build the composite figure using centimeter cubes.

9. Separate the figure into prisms. Make a drawing of each prism used to build the composite figure.

My Drawing!

10. How many cubes did it take to build the figure?

_____

11. What is the volume of this figure?

_____ cubic centimeters

Apply It

Tanela arranged centimeter cubes into the composite figure shown. Use the composite figure for Exercises 12 and 13.

12. Mathematical PRACTICE 4 Model Math Separate the figure into prisms. Make a drawing of each prism used to build the composite figure.

13. What is the volume of the composite figure? Check your answer by building a model and counting the number of cubes.

_____ cubic centimeters

14. Circle the composite figure that has a volume of 24 cubic centimeters.

15. Mathematical PRACTICE 1 Make Sense of Problems Explain how to use the formula of a rectangular prism to find the volume of a composite figure that is composed of rectangular prisms.

_____

_____

Write About It

16. How can you use models to find the volume of composite figures?

_____

_____

MY Homework

Homework Helper

Need help? connectED.mcgraw-hill.com

A composite figure is shown at the right. Centimeter cubes were used to build the figure. Find the volume.

1 Six cubes were used to make the base layer.

2 Four cubes were used to make the two top layers.

3 Add the number of cubes for the base and the top.
6 + 4 = 10

So, a total of 10 cubes were used to build the figure.
The volume is 10 cubic centimeters.

Practice

Refer to the composite figure at the right.

1. How many cubes are needed to build the bottom layer?

_____

2. How many cubes are needed to build the top two layers?

_____

3. Use addition to add the bottom and top layers.

_____

4. What is the volume of the composite figure?

_____ cubic centimeters

Problem Solving

Jared built the composite figure at the right using centimeter cubes.

5. Separate the figure into prisms. Make a drawing of each prism used to build the composite figure.

My Drawing!

6. How many cubes did it take Jared to build the figure?

_____

7. What is the volume of this figure? _____ cubic centimeters

8. **Mathematical** **PRACTICE** **Find the Error** Gabriele built a composite figure using 12 cubes for the bottom layer and 10 cubes for the top layer. She said that the volume of the composite figure was 12 × 10, or 120 cubic centimeters. Find and correct her error.

_____

_____

Vocabulary Check

Fill in the blank with the correct term or number to complete the sentence.

9. A composite figure is made up of two or more

_____ figures.

10. The composite figure was built using centimeter cubes. What is the volume of the composite figure shown?

$V =$ _____ cubic centimeters

Volume of Composite Figures

ESSENTIAL QUESTION
How does geometry help me solve problems in everyday life?

A **composite figure** is made up of two or more three-dimensional figures. To find the volume, separate the figure into figures with volumes you know how to find.

Math in My World

Tutor

Example 1

The Arc de Triomphe in Paris, France, is roughly in the shape of the composite figure shown. Find the volume of the Arc de Triomphe.

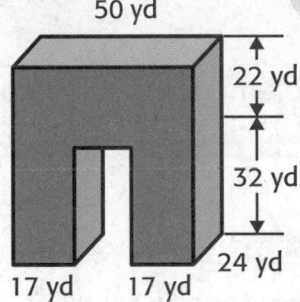

50 yd

22 yd

32 yd

24 yd

17 yd   17 yd

Separate the figure into three rectangular prisms. Find the volume of each prism.

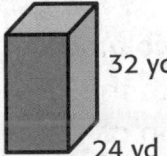

32 yd

24 yd

17 yd

$V = \ell \times w \times h$

$V = 17 \times 24 \times 32$

$V =$

32 yd

24 yd

17 yd

$V = \ell \times w \times h$

$V = 17 \times 24 \times 32$

$V =$

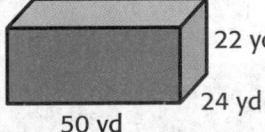

22 yd

24 yd

50 yd

$V = \ell \times w \times h$

$V = 50 \times 24 \times 22$

$V =$

$+$

So, the total volume is _____ cubic yards, or _____ yd³.

 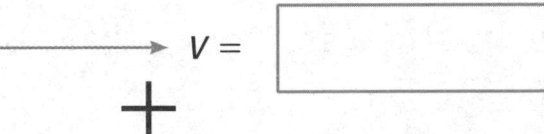

Example 2

Find the volume of the composite figure.

Separate the figure into two prisms.
Find the volume of each prism.

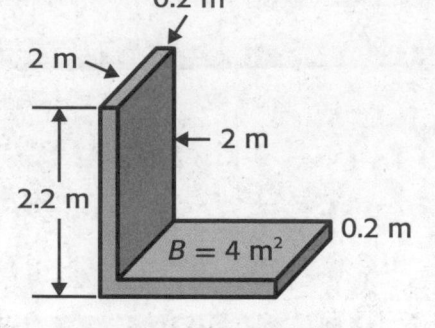

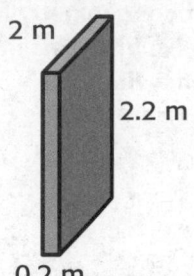

$$V = \ell \times w \times h$$

$$V = \underline{\hspace{1cm}} \times \underline{\hspace{1cm}} \times \underline{\hspace{1cm}} \longrightarrow V = \boxed{}$$

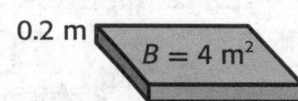

$$V = B \times h$$

$$V = \underline{\hspace{1cm}} \times \underline{\hspace{1cm}} \longrightarrow V = \boxed{}$$

$$+ \underline{\hspace{6cm}}$$

Add the volumes. The total volume is _____ cubic meters, or $\boxed{}$ m³.

Guided Practice ✓Check

1. Find the volume of the composite figure.

Bottom Prism

$$V = B \times h$$

$$V = 126 \times 11$$

$$V = \underline{\hspace{1.5cm}}$$

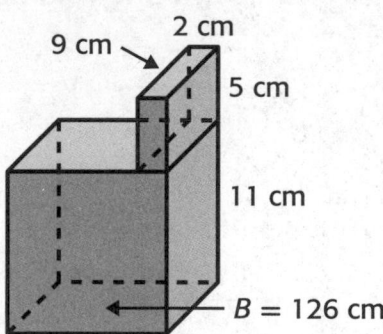

Top Prism

$$V = \ell \times w \times h$$

$$V = 2 \times 9 \times 5$$

$$V = 2 \times (9 \times 5) \qquad \text{Associative Property}$$

$$V = 2 \times 45$$

$$V = \underline{\hspace{1.5cm}}$$

The total volume is _____ + _____

or _____ cubic centimeters.

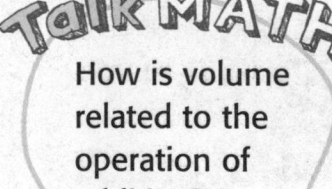

Talk MATH

How is volume related to the operation of addition?

Independent Practice

Find the volume of each composite figure.

2.

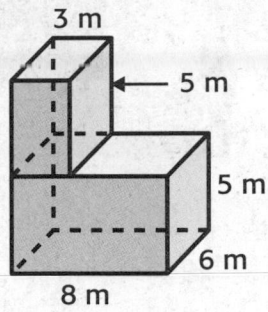

3 m
5 m
5 m
6 m
8 m

$V =$ _____

3.

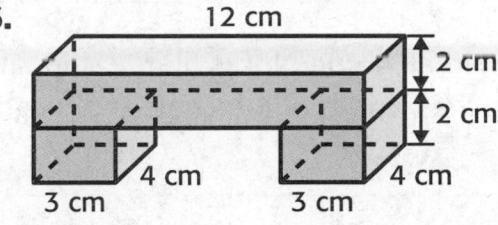

12 cm
2 cm
2 cm
3 cm   4 cm   3 cm   4 cm

$V =$ _____

4.

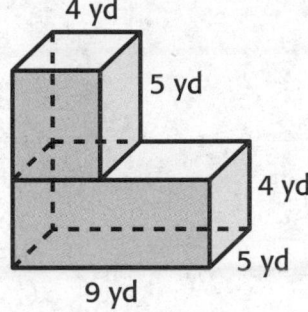

4 yd
5 yd
4 yd
5 yd
9 yd

$V =$ _____

5.

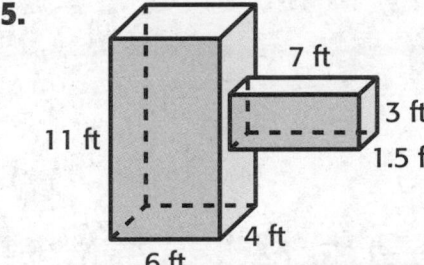

7 ft
3 ft
11 ft
1.5 ft
4 ft
6 ft

$V =$ _____

6.

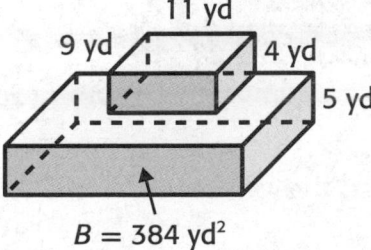

11 yd
9 yd   4 yd
5 yd
$B = 384$ yd^2

$V =$ _____

7.

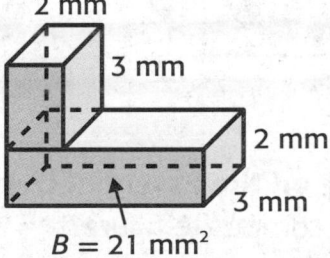

2 mm
3 mm
2 mm
3 mm
$B = 21$ mm^2

$V =$ _____

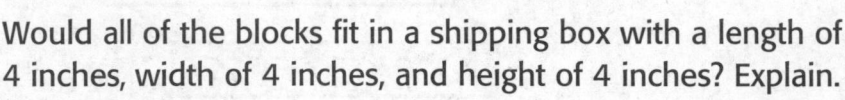
8. Mrs. Stafford ordered the set of blocks shown at the right for her classroom. Find the total volume of all the blocks.

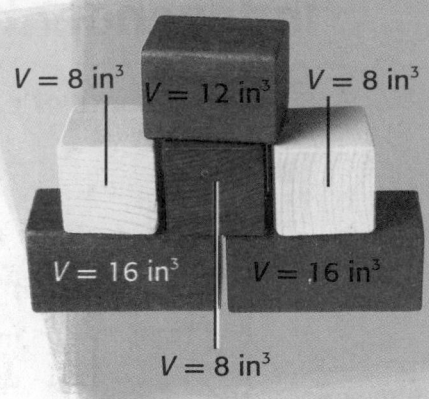

$V = 8 \text{ in}^3$ $V = 12 \text{ in}^3$ $V = 8 \text{ in}^3$

$V = 16 \text{ in}^3$ $V = 16 \text{ in}^3$

$V = 8 \text{ in}^3$

_____

Would all of the blocks fit in a shipping box with a length of 4 inches, width of 4 inches, and height of 4 inches? Explain.

_____

_____

_____

9. The figure represents a piece of foam packaging. Find the volume of the foam.

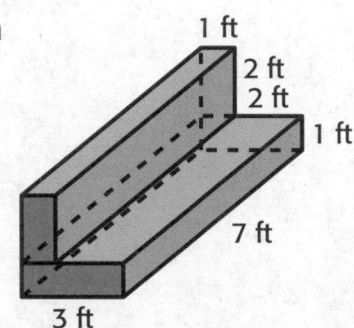

1 ft
2 ft
2 ft
1 ft
7 ft
3 ft

_____

HOT Problems

10. **Mathematical PRACTICE 3** **Which One Doesn't Belong?** Circle the figure that is not a composite figure.

      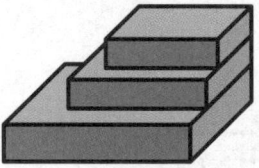

11. **Building on the Essential Question** How can I find the volume of a composite figure?

_____

_____

MY Homework

Homework Helper

Need help? connectED.mcgraw-hill.com

Find the volume of the composite figure.

The figure was separated into two prisms.
Find the volume of each rectangular prism.

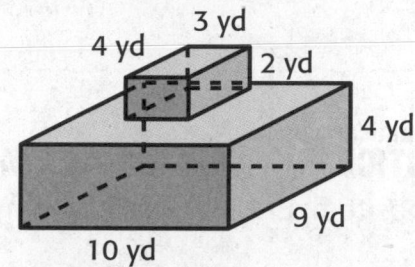

Top prism

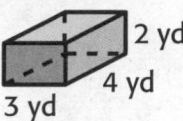

$V = \ell \times w \times h$

$V = 3 \times 4 \times 2 \longrightarrow 24$

Bottom prism

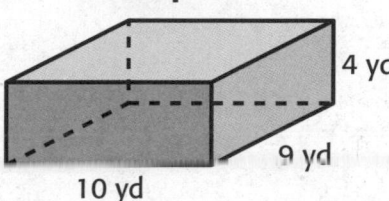

$V = \ell \times w \times h$

$V = 10 \times 9 \times 4 \longrightarrow \underline{+\ 360}$

384

The total volume of the composite figure is 24 + 360 or 384 cubic yards.

Practice

Find the volume of each composite figure.

1.

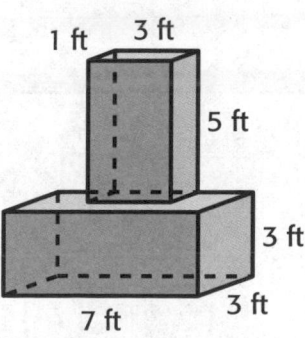

$V = $ _____

2.

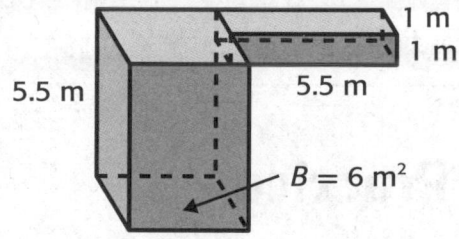

$V = $ _____

Problem Solving

3. Maci is decorating the cake shown. Find the volume of the cake.

_____

4. The firehouse shown is in the shape of a composite figure. How many cubic yards of space are in the firehouse?

_____

5. **Mathematical PRACTICE 4 Model Math** Draw an example of a composite figure that has a volume between 750 and 900 cubic units.

My Drawing!

Vocabulary Check

Fill in the blank with the correct term or number to complete the sentence.

6. A _____ is made up of two or more three-dimensional figures.

Test Practice

7. What is the total volume of the composite figure?

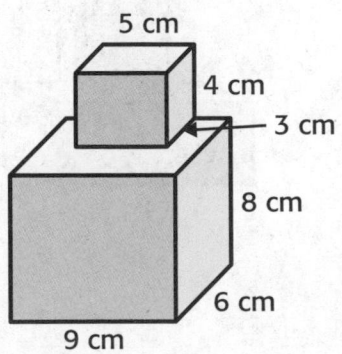

Ⓐ 282 cubic centimeters   Ⓒ 492 cubic centimeters

Ⓑ 432 cubic centimeters   Ⓓ 502 cubic centimeters

Problem-Solving Investigation
STRATEGY: Make a Model

Lesson 12

ESSENTIAL QUESTION
How does geometry help me solve problems in everyday life?

Learn the Strategy

Nick is helping his sister put away her alphabet blocks. To fill one layer, it takes nine blocks. If there are six layers, how many blocks would be in the box?

1 Understand

What facts do you know?

There are _____ blocks in each layer and there are six layers.

What do you need to find?

The number of blocks in the box when there are _____ layers.

2 Plan

I can solve the problem by making a _____ .

3 Solve

Arrange _____ cubes in a 3 × 3 array. Stack the cubes

until there are _____ layers. There are a total of

_____ cubes. So, the box would have _____ blocks.

4 Check

Is my answer reasonable? Explain.

Multiply.

6 × 9 = _____

Practice the Strategy

Evelyn wants to mail a package to her cousin. What is the volume of the package if it is 6 inches long, 4 inches wide, and 4 inches tall?

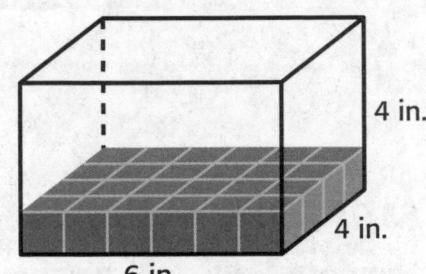

4 in.

4 in.

6 in.

① Understand

What facts do you know?

_____

_____

What do you need to find?

_____

② Plan

_____

③ Solve

④ Check

Is my answer reasonable? Explain.

_____

Apply the Strategy

Solve each problem by making a model.

1. On an assembly line that is 150 feet long, there is a work station every 15 feet. The first station is at the beginning of the line. How many work stations are there?

_____

2. **Mathematical PRACTICE** 5 **Use Math Tools** A store is stacking cans of food into a rectangular prism display. The bottom layer has 8 cans by 5 cans. There are 5 layers. How many cans are in the display?

_____

3. The distance around the center ring at the circus is 80 feet. A clown stands every 10 feet along the circle. How many clowns are there?

_____

4. Martino wants to arrange 18 square tiles into a rectangular shape with the least perimeter possible. Perimeter is the distance around a figure. How many tiles will be in each row?

_____

Review the Strategies

Use any strategy to solve each problem.

- Make a model.
- Guess, check, and revise.
- Look for a pattern.
- Make a table.

5. Five friends are standing in a circle and playing a game where they toss a ball of yarn to one another. If each person is connected by the yarn to each other person only once, how many lines of yarn will connect the group?

_____

6. **Mathematical PRACTICE 8** **Look for a Pattern** In the figure below, there are 22 marbles in Box A. To go from Box A to Box B, exactly four marbles must pass through the triangular machine at a time. Exactly five marbles must pass through the square machine at a time. Describe how to move all the marbles from Box A to Box B in the fewest moves possible.

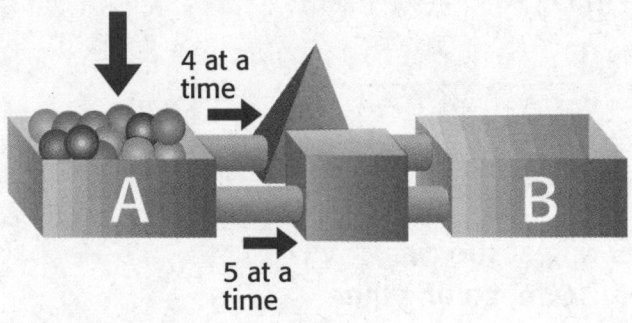

_____

_____

_____

7. The volume of a rectangular prism is 5,376 cubic inches. The prism is 14 inches long and 16 inches wide. How tall is the prism?

_____

8. The table at the right shows the number of minutes Danielle spent practicing the trumpet over the last 7 days. If she continues this pattern of practicing, in how many days will she have practiced 340 minutes?

Day	Time (min)
1	20
2	20
3	35
4	20
5	20
6	35
7	20

MY Homework

Homework Helper

Need help? connectED.mcgraw-hill.com

Victor wants to build a brick wall. Each brick layer is 3 inches thick, and the wall will be 18 inches tall. How many layers will it have?

1 Understand

What facts do you know?

Each brick layer is 3 inches thick. The wall will be 18 inches tall.

What do you need to find?

the number of layers the wall will have

2 Plan

Solve the problem by making a model.

3 Solve

Make a model of the wall by using cubes.

Each cube represents a 3-inch brick.

He needs 6 cubes to build the 18-inch wall.

So, the wall would have 6 layers.

```
      3 in.
    + 3 in.
    + 3 in.
    + 3 in.
    + 3 in.
    + 3 in.
    ───────
     18 in.
```

4 Check

Is my answer reasonable? Explain.

Multiply. $6 \times 3 = 18$ inches

![Real World] **Problem Solving**

Solve each problem by making a model.

1. Nan and Sato are designing a coffee table using 4-inch tiles. Nan uses 30 tiles and Sato uses half as many. How many total tiles did they use?

_____

If the area of the table is 36 inches by 24 inches, will they have enough tiles to cover the table? If not, how many more will they need?

_____

2. The Jones family is landscaping their yard. Their yard is 160 square feet and one side is 10 feet long. What is the length of the other side of the yard?

_____

If they plant 3 bushes that need to be 3 feet apart and 3 feet away from the fence around the yard, will they have the space?

_____

3. Billy is organizing his closet. He has clothing bins that measure 20 inches high, 18 inches wide, and 14 inches long. How many bins can he fit in a 60-inch long closet that is 36 inches deep and 72 inches high?

_____

4. **Mathematical PRACTICE** 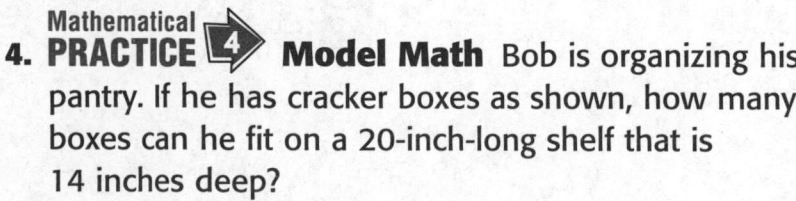 **Model Math** Bob is organizing his pantry. If he has cracker boxes as shown, how many boxes can he fit on a 20-inch-long shelf that is 14 inches deep?

_____

12 in.

10 in.   2 in.

Vocabulary Check

Match each word to its definition. Write your answers on the lines provided.

1. **equilateral triangle** _____

2. **composite figure** _____

3. **parallelogram** _____

4. **volume** _____

5. **rectangular prism** _____

6. **regular polygon** _____

7. **triangular prism** _____

8. **obtuse triangle** _____

9. **face** _____

10. **polygon** _____

11. **square** _____

12. **pentagon** _____

A. a three-dimensional figure with six rectangular faces, twelve edges, and eight vertices

B. a flat surface of a three-dimensional figure

C. a triangle with one obtuse angle

D. a closed figure made up of line segments that do not cross each other

E. a figure made up of two or more three-dimensional figures

F. a polygon with five sides

G. a prism with two congruent triangular bases

H. a polygon with congruent sides and all congruent angles

I. a quadrilateral with opposite sides both parallel and congruent

J. a triangle with three congruent sides

K. the amount of space within a three-dimensional figure

L. a rectangle with four congruent sides

Concept Check ✓

Name each polygon. Determine if it appears to be *regular* or *not regular*.

13.

_____

14.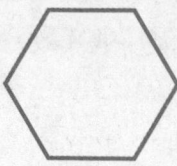

_____

Describe the attributes of each quadrilateral. Then classify the quadrilateral.

15.

_____

_____

16.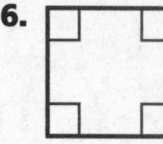

_____

_____

Describe the faces, edges, and vertices of each three-dimensional figure. Then identify it.

17.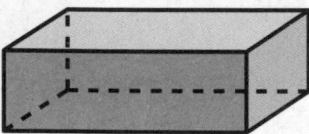

_____

_____

18.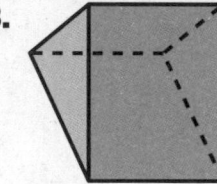

_____

_____

Problem Solving

19. A triangle forms the front of the Pantheon in Rome, Italy. Classify the triangle based on its sides. Then classify it based on its angles.

_____

_____

20. Shawn keeps his photos in a box like the one shown.

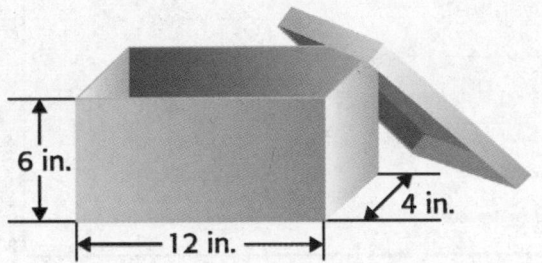

What is the volume of the box?

_____

21. Ishmail wants to build a long train track. If each piece of track is 6 inches long, and he has 42 pieces, can he make a track that is 20 feet long?

_____

Can he make a track that is 22 feet long?

_____

My Work!

Test Practice

22. Find the volume of the composite figure.

Ⓐ 2,700 in³   Ⓒ 3,420 in³

Ⓑ 2,780 in³   Ⓓ 3,660 in³

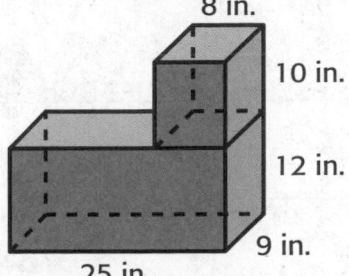

Reflect

Use what you learned about geometry to complete the
graphic organizer.

ESSENTIAL
QUESTION

How does geometry
help me solve problems
in everyday life?

Faces, Edges, and Vertices

Vocabulary

Real-World Example

Now reflect on the ESSENTIAL QUESTION Write your answer below.

_____

_____

Glossary/Glosario

Go online for the eGlossary.

Go to the *eGlossary* to find out more about these words in the following 13 languages:

Arabic • Bengali • Brazilian Portuguese • Cantonese • English • Haitian Creole
Hmong • Korean • Russian • Spanish • Tagalog • Urdu • Vietnamese

Aa	**English**	**Spanish/Español**

acute angle   An angle with a measure between 0° and 90°.

ángulo agudo   Ángulo que mide entre 0° y 90°.

acute triangle   A triangle with three acute angles.

triángulo acutángulo   Triángulo con tres ángulos agudos.

algebra   A branch of mathematics that uses symbols, usually letters, to explore relationships between quantities.

álgebra   Rama de las matemáticas que usa símbolos, generalmente letras, para explorar relaciones entre cantidades.

angle   Two rays with a common endpoint.

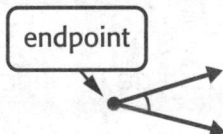
endpoint

ángulo   Dos semirrectas con un extremo común.

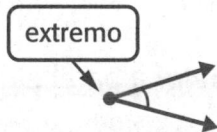
extremo

annex   To place a zero to the right of a decimal without changing a number's value.

agregar   Poner un cero a la derecha de un decimal sin cambiar el valor de un número.

area   The number of square units needed to cover the surface of a closed figure.

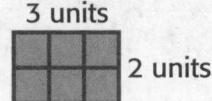

area = 6 square units

área   Cantidad de unidades cuadradas necesarias para cubrir la superficie de una figura cerrada.

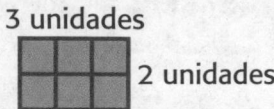

área = 6 unidades cuadradas

Associative Property   Property that states that the way in which numbers are grouped does not change the sum or product.

propiedad asociativa   Propiedad que establece que la manera en que se agrupan los números no altera la suma o el producto.

attribute   A characteristic of a figure.

atributo   Característica de una figura.

axis   A horizontal or vertical number line on a graph. Plural is axes.

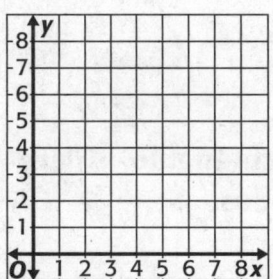

eje   Recta numérica horizontal o vertical en una gráfica.

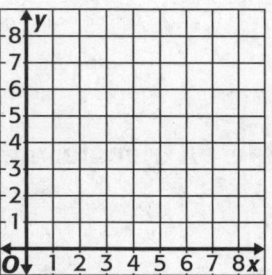

base   In a power, the number used as a factor. In 10^3, the base is 10.

base   En una potencia, el número que se usa como factor. En 10^3, la base es 10.

base   Any side of a parallelogram.

base   Cualquiera de los lados paralelogramo.

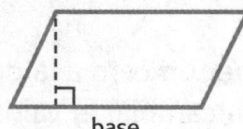

base

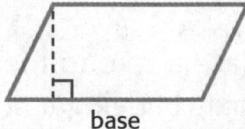

base

base   One of the two parallel congruent faces in a prism.

base   Una de las dos caras congruentes paralelas en un prisma.

capacity The amount a container can hold.

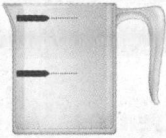

capacidad Cantidad que puede contener un recipiente.

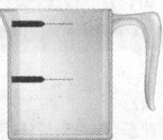

centimeter (cm) A metric unit for measuring length.

100 centimeters = 1 meter

centímetro (cm) Unidad métrica de longitud.

100 centímetros = 1 metro

common denominator A number that is a multiple of the denominators of two or more fractions.

denominador común Número que es múltiplo de los denominadores de dos o más fracciones.

common factor A number that is a factor of two or more numbers.

3 is a common factor of 6 and 12.

factor común Número que es un factor de dos o más números.

3 es factor común de 6 y 12.

common multiple A whole number that is a multiple of two or more numbers.

24 is a common multiple of 6 and 4.

múltiplo común Número natural múltiplo de dos o más números.

24 es un múltiplo común de 6 y 4.

Commutative Property Property that states that the order in which numbers are added does not change the sum and that the order in which factors are multiplied does not change the product.

propiedad conmutativa Propiedad que establece que el orden en que se suman los números no altera la suma y que el orden en que se multiplican los factores no altera el producto.

compatible numbers Numbers in a problem that are easy to work with mentally.

720 and 90 are compatible numbers for division because 72 ÷ 9 = 8.

números compatibles Números en un problema con los cuales es fácil trabajar mentalmente.

720 ÷ 90 es una divisíon que usa son números compatibles porque 72 ÷ 9 = 8.

composite figures A figure made up of two or more three-dimensional figures.

figura compuesta Figura conformada por dos o más figuras tridimensionales.

Cc

composite number A whole number that has more than two factors.

12 has the factors 1, 2, 3, 4, 6, and 12.

congruent Having the same measure.

congruent angles Angles of a figure that are equal in measure.

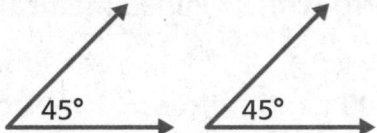

congruent figures Two figures having the same size and the same shape.

congruent sides Sides of a figure that are equal in length.

convert To change one unit to another.

coordinate One of two numbers in an ordered pair.

The 1 is the number on the *x*-axis, the 5 is on the *y*-axis.

coordinate plane A plane that is formed when two number lines intersect.

número compuesto Número natural que tiene más de dos factores.

12 tiene a los factores 1, 2, 3, 4, 6 y 12.

congruentes Que tienen la misma medida.

ángulos congruentes Ángulos de una figura que tienen la misma medida.

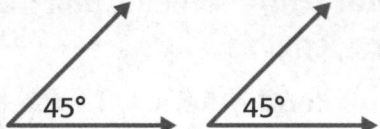

figuras congruentes Dos figuras que tienen el mismo tamaño y la misma forma.

lados congruentes Lados de una figura que tienen la misma longitud.

convertir Transformar una unidad en otra.

coordenada Cada uno de los números de un par ordenado.

El 1 es la coordenada *x* y el 5 es la coordenada *y*.

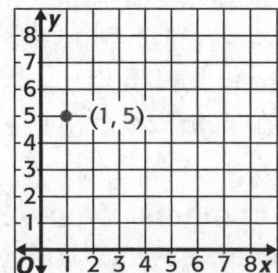

plano de coordenadas Plano que se forma cuando dos rectas numéricas se intersecan formando un ángulo recto.

cube A rectangular prism with six faces that are congruent squares.

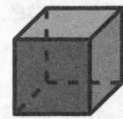

cubed A number raised to the third power; $10 \times 10 \times 10$, or 10^3.

cubic unit A unit for measuring volume, such as a cubic inch or a cubic centimeter.

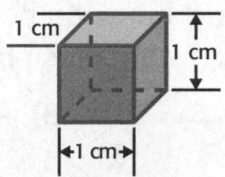

cup A customary unit of capacity equal to 8 fluid ounces.

customary system The units of measurement most often used in the United States. These include foot, pound, and quart.

cubo Prisma rectangular con seis caras que son cuadrados congruentes.

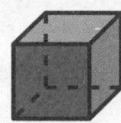

al cubo Número elevado a la tercera potencia; $10 \times 10 \times 10$ o 10^3.

unidad cúbica Unidad de volumen, como una pulgada cúbica o un centímetro cúbico.

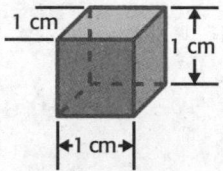

taza Unidad usual de capacidad que equivale a 8 onzas líquidas.

sistema usual Conjunto de unidades de medida de uso más frecuente en Estados Unidos. Incluyen el pie, la libra y el cuarto.

Dd

decimal A number that has a digit in the tenths place, hundredths place, and beyond.

decimal point A period separating the ones and the tenths in a decimal number.

0.8 or $3.77

degree (°) a. A unit of measure used to describe temperature. b. A unit for measuring angles.

decimal Número que tiene al menos un dígito en el lugar de las décimas, centésimas etcétera.

punto decimal Punto que separa las unidades y las décimas en un número decimal.

0.8 o $3.77

grado (°) a. Unidad de medida que se usa para describir la temperatura. b. Unidad que se usa para medir ángulos.

Dd

denominator The bottom number in a fraction. It represents the number of parts in the whole.

In $\frac{5}{6}$, 6 is the denominator.

digit A symbol used to write numbers. The ten digits are 0, 1, 2, 3, 4, 5, 6, 7, 8, and 9.

Distributive Property To multiply a sum by a number, you can multiply each addend by the same number and add the products.

$$8 \times (9 + 5) = (8 \times 9) + (8 \times 5)$$

divide (division) An operation on two numbers in which the first number is split into the same number of equal groups as the second number.

12 ÷ 3 means 12 is divided into 3 equal-size groups

dividend A number that is being divided.

$3\overline{)19}$ ← 19 is the dividend.

divisible Describes a number that can be divided into equal parts and has no remainder.

39 is divisible by 3 with no remainder.

divisor The number that divides the dividend.

3 is the divisor. → $3\overline{)19}$

---

denominador Numero que se escribe debajo de la barra en una fracción. Representa el número de partes en que se divide un entero.

En $\frac{5}{6}$, 6 es el denominador.

dígito Símbolo que se usa para escribir los números. Los diez dígitos son 0, 1, 2, 3, 4, 5, 6, 7, 8 y 9.

propiedad distributiva Para multiplicar una suma por un número, puedes multiplicar cada sumando por ese número y luego sumar los productos.

$$8 \times (9 + 5) = (8 \times 9) + (8 \times 5)$$

dividir (división) Operación entre dos números en la cual el primer número se separa en tantos grupos iguales como indica el segundo número.

12 ÷ 3 significa que 12 se divide entre 3 grupos de igual tamaño.

dividendo Número que se divide.

$3\overline{)19}$ ← 19 es el dividendo.

divisible Describe un número que puede dividirse en partes iguales, sin residuo.

39 es divisible entre 3 sin residuo.

divisor Número entre el cual se divide el dividendo.

3 es el divisor. → $3\overline{)19}$

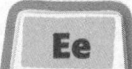

edge The line segment where two faces of a three-dimensional figure meet.

arista Segmento de recta donde se unen dos caras de una figura tridimensional.

equation A number sentence that contains an equal sign, showing that two expressions are equal.

ecuación Expresión numérica que contiene un signo igual y que muestra que dos expresiones son iguales.

equilateral triangle A *triangle* with three *congruent* sides.

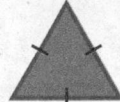

triángulo equilátero *Triángulo* con tres lados *congruentes*.

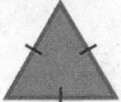

equivalent decimals Decimals that have the same value.

0.3 and 0.30

decimales equivalentes Decimales que tienen el mismo valor.

0.3 y 0.30

equivalent fractions Fractions that have the same value.

$$\frac{3}{4} = \frac{6}{8} = \frac{9}{12}$$

$\frac{1}{4}$	$\frac{1}{4}$	$\frac{1}{4}$		
$\frac{1}{8}$ $\frac{1}{8}$ $\frac{1}{8}$ $\frac{1}{8}$ $\frac{1}{8}$ $\frac{1}{8}$			
$\frac{1}{12}$ $\frac{1}{12}$ $\frac{1}{12}$ $\frac{1}{12}$ $\frac{1}{12}$ $\frac{1}{12}$ $\frac{1}{12}$ $\frac{1}{12}$ $\frac{1}{12}$			

fracciones equivalentes Fracciones que tienen el mismo valor.

$$\frac{3}{4} = \frac{6}{8} = \frac{9}{12}$$

$\frac{1}{4}$	$\frac{1}{4}$	$\frac{1}{4}$		
$\frac{1}{8}$ $\frac{1}{8}$ $\frac{1}{8}$ $\frac{1}{8}$ $\frac{1}{8}$ $\frac{1}{8}$			
$\frac{1}{12}$ $\frac{1}{12}$ $\frac{1}{12}$ $\frac{1}{12}$ $\frac{1}{12}$ $\frac{1}{12}$ $\frac{1}{12}$ $\frac{1}{12}$ $\frac{1}{12}$			

estimate A number close to an exact value. An estimate indicates about how much.

47 + 22 (round to 50 + 20)

The estimate is 70.

estimación Número cercano a un valor exacto. Una estimación indica una cantidad aproximada.

47 + 22 (se redondea a 50 + 20)

La estimación es 70.

evaluate   To find the value of an expression by replacing variables with numbers.

even number   A whole number that is divisible by 2.

expanded form   A way of writing a number as the sum of the values of its digits.

exponent   In a power, the number of times the base is used as a factor. In 5^3, the exponent is 3.

expression   A combination of numbers, variables, and at least one operation.

evaluar   Calcular el valor de una expresión reemplazando las variables por números.

número par   Número natural divisible entre 2.

forma desarrollada   Manera de escribir un número como la suma de los valores de sus dígitos.

exponente   En una potencia, el número de veces que se usa la base como factor. En 5^3, el exponente es 3.

expresión   Combinación de números, variables y por lo menos una operación.

Ff

face   A flat surface.

A square is a face of a cube.

cara   Superficia plana.

Cada cara de un cubo es un cuadrado.

fact family   A group of related facts using the same numbers.

factor   A number that is multiplied by another number.

Fahrenheit (°F)   A unit used to measure temperature.

fair share   An amount divided equally.

fluid ounce   A customary unit of capacity.

familia de operaciones   Grupo de operaciones relacionadas que usan los mismos números.

factor   Número que se multiplica por otro número.

Fahrenheit (°F)   Unidad que se usa para medir la temperatura.

partes iguales   Partes entre las que se divide equitativamente un entero.

onza líquida   Unidad usual de capacidad.

foot (ft) A customary unit for measuring length. Plural is feet.

1 foot = 12 inches

fraction A number that represents part of a whole or part of a set.

$$\frac{1}{2}, \frac{1}{3}, \frac{1}{4}, \frac{3}{4}$$

pie (pie) Unidad usual de longitud.

1 pie = 12 pulgadas

fracción Número que representa una parte de un todo o una parte de un conjunto.

$$\frac{1}{2}, \frac{1}{3}, \frac{1}{4}, \frac{3}{4}$$

gallon (gal) A customary unit for measuring capacity for liquids.

1 gallon = 4 quarts

galón (gal) Unidad de medida usual de capacidad de líquidos.

1 galón = 4 cuartos

gram (g) A metric unit for measuring mass.

gramo (g) Unidad métrica para medir la masa.

graph To place a point named by an ordered pair on a coordinate plane.

graficar Colocar un punto nombrado por un par ordenado en un plano de coordenadas.

Greatest Common Factor (GCF) The greatest of the common factors of two or more numbers.

The greatest common factor of 12, 18, and 30 is 6.

máximo común divisor (M.C.D.) El mayor de los factores comunes de dos o más números.

El máximo común divisor de 12, 18 y 30 es 6.

height The shortest distance from the base of a parallelogram to its opposite side.

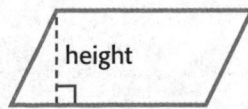
height

altura La distancia más corta desde la base de un paralelogramo hasta su lado opuesto.

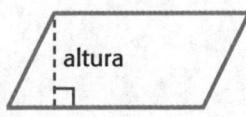
altura

Hh

hexagon   A polygon with six sides and six angles.

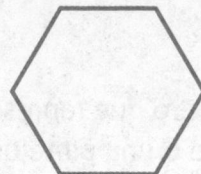

horizontal axis   The axis in a coordinate plane that runs left and right (↔). Also known as the *x*-axis.

hundredth   A place value position. One of one hundred equal parts. In the number 0.57, 7 is in the hundredths place.

hexágono   Polígono con seis lados y seis ángulos.

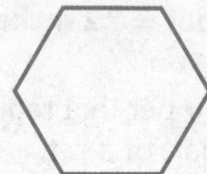

eje horizontal   Eje en un plano de coordenadas que va de izquierda a derecha (↔). También conocido como eje *x*.

centésima   Valor posícional. Una de cien partes iguales. En el número 0.57, 7 está en el lugar de las centésimas.

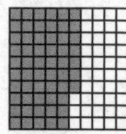

Ii

Identity Property   Property that states that the sum of any number and 0 equals the number and that the product of any number and 1 equals the number.

propiedad de identidad   Propiedad que establece que la suma de cualquier número y 0 es igual al número y que el producto de cualquier número y 1 es igual al número.

improper fraction   A fraction with a numerator that is greater than or equal to the denominator.

$$\frac{17}{3} \text{ or } \frac{5}{5}$$

fracción impropia   Fracción con un numerador mayor que él igual al denominador.

$$\frac{17}{3} \text{ o } \frac{5}{5}$$

inch (in.)   A customary unit for measuring length. The plural is inches.

pulgada (pulg)   Unidad usual de longitud.

inequality   Two quantities that are not equal.

desigualdad   Dos cantidades que no son iguales.

intersecting lines *Lines* that meet or cross at a common *point*.

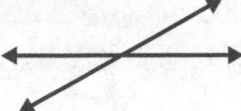

rectas secantes *Rectas* que se intersecan o se cruzan en un *punto* común.

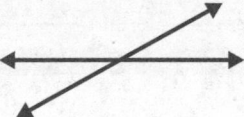

interval The distance between successive values on a scale.

intervalo Distancia entre valores sucesivos en una escala.

inverse operations Operations that undo each other.

operaciones inversas Operaciones que se cancelan entre sí.

isosceles triangle A *triangle* with at least 2 *sides* of the same *length*.

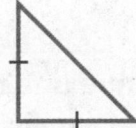

triángulo isósceles *Triángulo* que tiene por lo menos 2 *lados* del mismo largo.

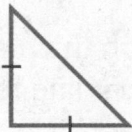

Kk

kilogram (kg) A metric unit for measuring mass.

kilogramo (kg) Unidad métrica de masa.

kilometer (km) A metric unit for measuring length.

kilómetro (km) Unidad métrica de longitud.

Ll

Least Common Denominator (LCD) The least common multiple of the denominators of two or more fractions.

$$\frac{1}{12}, \frac{1}{6}, \frac{1}{8}; \text{ LCD is 24.}$$

mínimo común denominador (m.c.d.) El mínimo común múltiplo de los denominadores de dos o más fracciones.

$$\frac{1}{12}, \frac{1}{6}, \frac{1}{8}; \text{ el m.c.d. es 24.}$$

Least Common Multiple (LCM) The smallest whole number greater than 0 that is a common multiple of each of two or more numbers.

The LCM of 2 and 3 is 6.

mínimo común múltiplo (m.c.m.) El menor número natural, mayor que 0, múltiplo común de dos o más números.

El m.c.m. de 2 y 3 es 6.

Ll

length Measurement of the distance between two points.

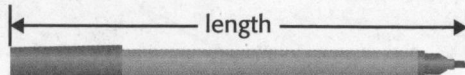

longitud Medida de la distancia entre dos puntos.

like fractions Fractions that have the same denominator.

$$\frac{1}{5} \text{ and } \frac{2}{5}$$

fracciones semejantes Fracciones que tienen el mismo denominador.

$$\frac{1}{5} \text{ y } \frac{2}{5}$$

line A set of *points* that form a straight path that goes on forever in opposite directions.

recta Conjunto de *puntos* que forman una trayectoria recta sin fin en direcciones opuestas.

line plot A graph that uses columns of Xs above a number line to show frequency of data.

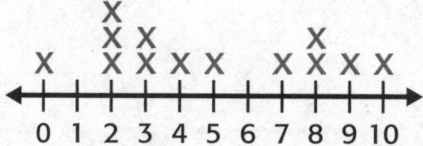

diagrama lineal Gráfica que usa columnas de X sobre una recta numérica para mostrar la frecuencia de los datos.

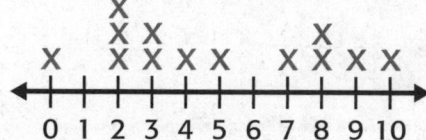

line segment A part of a *line* that connects two points.

segmento de recta Parte de una *recta* que conecta dos puntos.

liter (L) A metric unit for measuring volume or capacity.

1 liter = 1,000 milliliters

litro (L) Unidad métrica de volumen o capacidad.

1 litro = 1,000 mililitros

mass Measure of the amount of matter in an object.

masa Medida de la cantidad de materia en un cuerpo.

meter (m) A metric unit used to measure length.

metro (m) Unidad métrica que se usa para medir la longitud.

metric system (SI) The decimal system of measurement. Includes units such as meter, gram, and liter.

sistema métrico (SI) Sistema decimal de medidas que incluye unidades como el metro, el gramo y el litro.

mile (mi) A customary unit of measure for length.

$$1 \text{ mile} = 5{,}280 \text{ feet}$$

milla (mi) Unidad usual de longitud.

$$1 \text{ milla} = 5{,}280 \text{ pies}$$

milligram (mg) A metric unit used to measure mass.

$$1{,}000 \text{ milligrams} = 1 \text{ gram}$$

miligramo (mg) Unidad métrica de masa.

$$1{,}000 \text{ miligramos} = 1 \text{ gramo}$$

milliliter (mL) A metric unit used for measuring capacity.

$$1{,}000 \text{ milliliters} = 1 \text{ liter}$$

mililitro (mL) Unidad métrica de capacidad.

$$1{,}000 \text{ mililitros} = 1 \text{ litro}$$

millimeter (mm) A metric unit used for measuring length.

$$1{,}000 \text{ millimeters} = 1 \text{ meter}$$

milímetro (mm) Unidad métrica de longitud.

$$1{,}000 \text{ milímetros} = 1 \text{ metro}$$

mixed number A number that has a whole number part and a fraction part. $3\frac{1}{2}$ is a mixed number.

número mixto Número formado por un número natural y una parte fraccionaria. $3\frac{1}{2}$ es un número mixto.

multiple (multiples) A multiple of a number is the product of that number and any whole number.

15 is a multiple of 5 because
$$3 \times 5 = 15.$$

múltiplo Un múltiplo de un número es el producto de ese número por cualquier otro número natural.

15 es múltiplo de 5 porque
$$3 \times 5 = 15.$$

multiplication An operation on two numbers to find their product. It can be thought of as repeated addition. 4×3 is another way to write the sum of four 3s, which is $3 + 3 + 3 + 3$ or 12.

multiplicación Operación entre dos números para hallar su producto. También se puede interpretar como una suma repetida. 4×3 es otra forma de escribir la suma de cuatro veces 3, la cual es $3 + 3 + 3 + 3$ o 12.

net   A two-dimensional pattern of a three-dimensional figure.

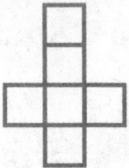

modelo plano   Patrón bidimensional de una figura tridimensional.

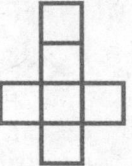

number line   A line that represents numbers as points.

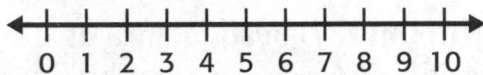

recta numérica   Recta que representa números como puntos.

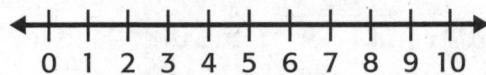

numerator   The top number in a fraction; the part of the fraction that tells the number of parts you have.

numerador   Número que se escribe sobre la barra de fracción; la parte de la fracción que indica el número de partes que hay.

numerical expression   A combination of numbers and operations.

expresión numérica   Combinación de números y operaciones.

obtuse angle   An angle that measures between 90° and 180°.

ángulo obtuso   Ángulo que mide entre 90° y 180°.

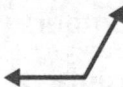

obtuse triangle   A *triangle* with one *obtuse angle.*

triángulo obtusángulo   *Triángulo* con un *ángulo obtuso.*

octagon   A polygon with eight sides.

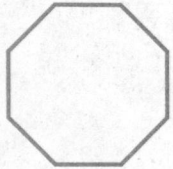

octágono   Polígono de ocho lados.

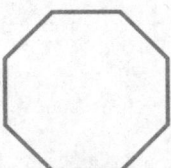

odd number   A number that is not divisible by 2; such a number has 1, 3, 5, 7, or 9 in the ones place.

número impar   Número que no es divisible entre 2. Los números impares tienen 1, 3, 5, 7, o 9 en el lugar de las unidades.

order of operations   A set of rules to follow when more than one operation is used in an expression.
   1. Perform operations in parentheses.
   2. Find the value of exponents.
   3. Multiply and divide in order from left to right.
   4. Add and subtract in order from left to right.

orden de las operaciones   Conjunto de reglas a seguir cuando se usa más de una operación en una expresión.
   1. Realiza las operaciones dentro de los paréntesis.
   2. Halla el valor de las potencias.
   3. Multiplica y divide de izquierda a derecha.
   4. Suma y resta de izquierda a derecha.

ordered pair   A pair of numbers that is used to name a point on the coordinate plane.

par ordenado   Par de números que se usa para nombrar un punto en un plano de coordenadas.

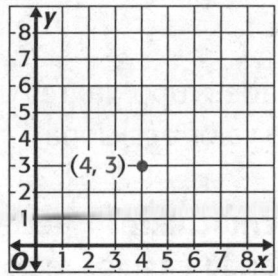

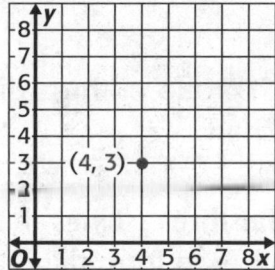

origin   The point (0, 0) on a coordinate plane where the vertical axis meets the horizontal axis.

origen   El punto (0, 0) en un plano de coordenadas donde el eje vertical interseca el eje horizontal.

ounce (oz)   A customary unit for measuring weight or capacity.

onza (oz)   Unidad usual de peso o capacidad.

parallel lines   Lines that are the same distance apart. Parallel lines do not meet.

rectas paralelas   Rectas separadas por la misma distancia en cualquier punto. Las rectas paralelas no se intersecan.

Pp

parallelogram   A quadrilateral with four sides in which each pair of opposite sides are parallel and congruent.

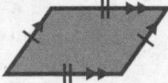

paralelogramo   Cuadrilátero en el cual cada par de lados opuestos son paralelos y congruentes.

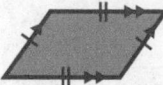

partial quotients   A method of dividing where you break the dividend into sections that are easy to divide.

cocientes parciales   Método de división por el cual se descompone el dividendo en secciones que son fáciles de dividir.

pentagon   A polygon with five sides.

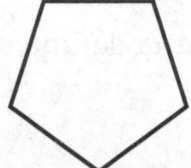

pentágono   Polígono de cinco lados.

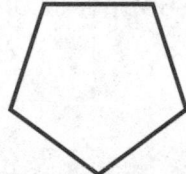

perimeter   The *distance* around a polygon.

perímetro   *Distancia* alrededor de un polígono.

period   Each group of three digits on a place-value chart.

período   Cada grupo de tres dígitos en una tabla de valor posicional.

perpendicular lines   Lines that meet or cross each other to form right angles.

rectas perpendiculares   Rectas que se cruzan formando ángulos rectos.

pint (pt)   A customary unit for measuring capacity.

1 pint = 2 cups

pinta (pt)   Unidad usual de capacidad.

1 pinta = 2 tazas

place   The position of a digit in a number.

posición   Lugar que ocupa un dígito en un número.

place value   The value given to a digit by its position in a number.

valor posicional   Valor dado a un dígito según su posición en el número.

place-value chart   A chart that shows the value of the digits in a number.

tabla de valor posicional   Tabla que muestra el valor de los dígitos en un número.

plane A flat surface that goes on forever in all directions.

point An exact location in space that is represented by a dot.

polygon A closed figure made up of line segments that do not cross each other.

positive number Number greater than zero.

pound (lb) A customary unit for measuring weight or mass.

1 pound = 16 ounces

power A number obtained by raising a base to an exponent.

$5^2 = 25$ **25 is a power of 5.**

power of 10 A number like 10, 100, 1,000 and so on. It is the result of using only 10 as a factor.

prime factorization A way of expressing a composite number as a product of its prime factors.

prime number A whole number with exactly two factors, 1 and itself.

7, 13, and 19

prism A three-dimensional figure with two parallel, congruent faces, called bases. At least three faces are rectangles.

product The answer to a multiplication problem.

plano Superficie plana que se extiende infinitamente en todas direcciones.

punto Ubicación exacta en el espacio que se representa con una marca puntual.

polígono Figura cerrada compuesta por segmentos de recta que no se intersecan.

número positivo Número mayor que cero.

libra (lb) Unidad usual de peso.

1 libra = 16 onzas

potencia Número que se obtiene elevando una base a un exponente.

$5^2 = 25$ **25 es una potencia de 5.**

potencia de 10 Número como 10, 100, 1,000, etc. Es el resultado de solo usar 10 como factor.

factorización prima Manera de escribir un número compuesto como el producto de sus factores primos.

número primo Número natural que tiene exactamente dos factores: 1 y sí mismo.

7, 13 y 19

prisma Figura tridimensional con dos caras congruentes y paralelas llamadas bases. Al menos tres caras son rectangulares.

producto Repuesta a un problema de multiplicación.

Pp

proper fraction   A fraction in which the numerator is less than the denominator.

$$\frac{1}{2}$$

fracción propia   Fracción en la que el numerador es menor que el denominador.

$$\frac{1}{2}$$

property   A rule in mathematics that can be applied to all numbers.

propiedad   Regla de las matemáticas que puede aplicarse a todos los números.

protractor   A tool used to measure and draw angles.

transportador   Instrumento que se usa para medir y trazar ángulos.

quadrilateral   A polygon that has 4 sides and 4 angles.

square, rectangle, parallelogram, and trapezoid

cuadrilátero   Polígono con 4 lados y 4 ángulos.

cuadrado, rectángulo, paralelogramo y trapezoide

quart (qt)   A customary unit for measuring capacity.

1 quart = 4 cups

cuarto (ct)   Unidad usual de capacidad.

1 cuarto = 4 tazas

quotient   The result of a division problem.

cociente   Resultado de un problema de división.

ray   A line that has one endpoint and goes on forever in only one direction.

semirrecta   Parta de una recta que tiene un extremo que se extiende infinitamente en una sola dirección.

rectangle   A quadrilateral with four right angles; opposite sides are equal and parallel.

rectángulo   Cuadrilátero con cuatro ángulo rectos; los lados opuestos son iguales y paralelos.

rectangular prism   A prism that has six rectangular bases.

prisma rectangular   Prisma que tiene bases rectangulares.

regular polygon   A polygon in which all sides are congruent and all angles are congruent.

polígono regular   Polígono que tiene todos los lados congruentes y todos los ángulos congruentes.

remainder   The number that is left after one whole number is divided by another.

residuo   Número que queda después de dividir un número natural entre otro.

rhombus   A *parallelogram* with four *congruent sides*.

rombo   *Paralelogramo* con cuatro *lados congruentes*.

right angle   An angle with a measure of 90°.

ángulo recto   Ángulo que mide 90°.

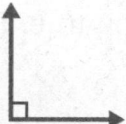

right triangle   A *triangle* with one *right angle*.

triángulo rectángulo   *Triángulo* con un *ángulo recto*.

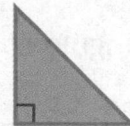

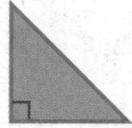

rounding   To find the approximate value of a number.

redondear   Hallar el valor aproximado de un número.

6.38 rounded to the nearest tenth is 6.4.

6.38 redondeado a la décima más cercana es 6.4.

scale   A set of numbers that includes the least and greatest values separated by equal intervals.

escala   Conjunto de números que incluye los valores menor y mayor separados por intervalos iguales.

scalene triangle   A *triangle* with no *congruent sides*.

triángulo escaleno   *Triángulo* sin *lados congruentes*.

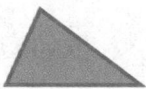

Ss

scaling   The process of resizing a number when it is multiplied by a fraction that is greater than or less than 1.

simplificar   Proceso de redimensionar un número cuando se multiplica por una fracción que es mayor que o menor que 1.

sequence   A list of numbers that follow a specific pattern.

secuencia   Lista de números que sigue un patrón específico.

simplest form   A fraction in which the GCF of the numerator and the denominator is 1.

forma simplificada   Fracción en la cual el M.C.D. del numerador y del denominador es 1.

solution   The value of a variable that makes an equation true. The solution of $12 = x + 7$ is 5.

solución   Valor de una variable que hace que la ecualción sea verdadera. La solución de $12 = x + 7$ es 5.

solve   To replace a variable with a value that results in a true sentence.

resolver   Remplazar una variable por un valor que hace que la expresión sea verdadera.

square   A rectangle with four *congruent sides*.

cuadrado   Rectángulo con cuatro *lados congruentes*.

square number   A number with two identical factors.

número al cuadrado   Número con dos factores idénticos.

square unit   A unit for measuring area, such as square inch or square centimeter.

unidad cuadrada   Unidad de área, como una pulgada cuadrada o un centímetro cuadrado.

squared   A number raised to the second power; 3×3, or 3^2.

al cuadrado   Número elevado a la segunda potencia; 3×3 o 3^2.

standard form   The usual or common way to write a number using digits.

forma estándar   Manera usual o común de escribir un número usando dígitos.

straight angle   An angle with a measure of 180°.

ángulo llano   Ángulo que mide 180°.

sum The answer to an addition problem.

suma Respuesta que se obtiene al sumar.

tenth A place value in a decimal number or one of ten equal parts or $\frac{1}{10}$.

décima Valor posicional en un número decimal o una de diez partes iguales o $\frac{1}{10}$.

term A number in a pattern or sequence.

término Cada número en un patrón o una secuencia.

thousandth(s) One of a thousand equal parts or $\frac{1}{1,000}$. Also refers to a place value in a decimal number. In the decimal 0.789, the 9 is in the thousandths place.

milésima(s) Una de mil partes iguales o $\frac{1}{1,000}$. También se refiere a un valor posicional en un número decimal. En el decimal 0.789, el 9 está en el lugar de las milésimas.

three-dimensional figure A figure that has length, width, and height.

figura tridimensional Figura que tiene largo, ancho y alto.

ton (T) A customary unit for measuring weight. 1 ton = 2,000 pounds

tonelada (T) Unidad usual de peso. 1 tonelada = 2,000 libras

trapezoid A quadrilateral with exactly one pair of parallel sides.

trapecio Cuadrilátero con exactamente un par de lados paralelos.

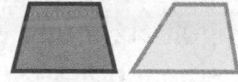

triangle A polygon with three sides and three angles.

triángulo Polígono con tres lados y tres ángulos.

triangular prism A prism that has triangular bases.

prisma triangular Prisma con bases triangulares.

unit cube   A cube with a side length of one unit.

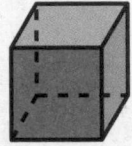

cubo unitario   Cubo con lados de una unidad de longitud.

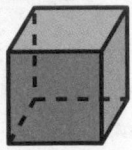

unit fraction   A fraction with 1 as its numerator.

fracción unitaria   Fracción que tiene 1 como su numerador.

unknown   A missing value in a number sentence or equation.

incógnita   Valor que falta en una oración numérica o una ecuación.

unlike fractions   Fractions that have different denominators.

fracciones no semejantes   Fracciones que tienen denominadores diferentes.

Vv

variable   A letter or symbol used to represent an unknown quantity.

variable   Letra o símbolo que se usa para representar una cantidad desconocida.

vertex   The point where two rays meet in an angle or where three or more faces meet on a three-dimensional figure.

vértice   a. Punto donde se unen los dos lados de un ángulo. b. Punto en una figura tridimensional donde se intersecan 3 o más aristas.

vertical axis   A vertical number line on a graph ($\updownarrow$). Also known as the *y*-axis.

eje vertical   Recta numérica vertical en una gráfica ($\updownarrow$). También conocido como eje *y*.

volume   The amount of space inside a three-dimensional figure.

volumen   Cantidad de espacio que contiene una figura tridimensional.

weight A measurement that tells how heavy an object is.

peso Medida que indica cuán pesado o liviano es un cuerpo.

x-axis The horizontal axis (↔) in a coordinate plane.

eje x Eje horizontal (↔) en un plano de coordenadas.

x-coordinate The first part of an ordered pair that indicates how far to the right of the *y*-axis the corresponding point is.

coordenada x Primera parte de un par ordenado; indica a qué distancia a la derecha del eje *y* está el punto correspondiente.

yard (yd) A customary unit of length equal to 3 feet or 36 inches.

yarda (yd) Unidad usual de longitud igual a 3 pies o 36 pulgadas.

y-axis The vertical axis (↕) in a coordinate plane.

eje y Eje vertical (↕) en un plano de coordenadas.

y-coordinate The second part of an ordered pair that indicates how far above the *x*-axis the corresponding point is.

coordenada y Segunda parte de un par ordenado; indica a qué distancia por encima del eje *x* está el punto correspondiente.

Name

Work Mat 1: Number Lines

0 1 2 3 4 5 6 7 8 9 10

0 1 2 3 4 5 6 7 8 9 10 11 12 13 14 15 16 17 18 19 20

Work Mat 3: Centimeter Grid

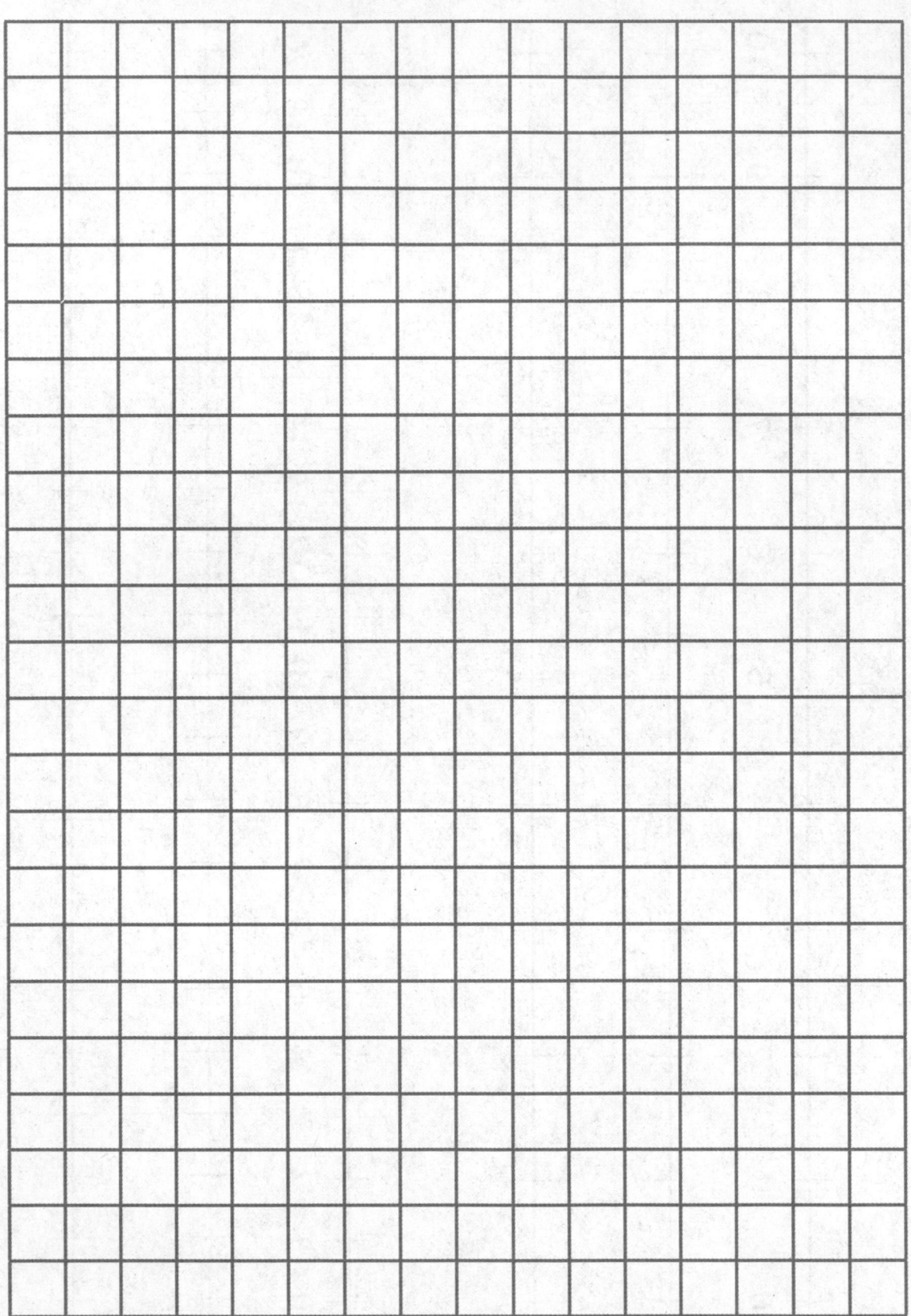

Name ....

Work Mat 4: Place-Value Chart (Hundreds to Thousandths)

Decimals			Ones		
thousandths	hundredths	tenths	ones	tens	hundreds

Work Mat 5: Tenths and Hundredths Models

Work Mat 6: Algebra Mat

Work Mat 8: First-Quadrant Grid (blank)